Penny Ante Science, Revisited

ISBN-13: 978-1720566496
ISBN-10: 1720566496

Notes to Teachers

Long ago Fred Fifer and Cynthia Ledbetter were faced with the issue of teaching hands-on science with little or no equipment and no budget with which to purchase supplies. Independently, they began writing their own labs, adapting what was in textbooks, and building, borrowing, begging, or buying what they could to be able to teach science. Years later they met at The University of Texas at Dallas and the *Penny Ante Science* series was born. The activities in this book are designed to be used with any type of student. The authors have resisted the temptation to assign grade levels because a lesson that appears appropriate for young children may be used as a lesson introduction for students in advanced classes. More complicated lessons can be simplified for less sophisticated children; to do this you may have to change the difficulty of questioning, or read the directions to the children. The authors encourage you to adapt these activities as you see fit, but to ensure that they remain discovery rather than confirmatory type activities.

While activities are divided into the traditional science sections, they are also indexed in a manner that allows you to integrate the sciences. For instance, you may use *Clonin' Jet Liners* as a means to teach observation skills, physical properties of matter, or as a focus for a study of the aerodynamics. *High Water, Low Water* may be an earth science, an environmental science, or a physics activity.

Several of the activities in this book use potentially dangerous equipment (hot plates, boiling water, open flames, etc.). Safety equipment is listed and directions for use are given. **Please explain all safety procedures to your students prior to doing the activities; require them to follow these procedures. Please review and follow all safety procedures prior to doing any preparation required of the teacher.** The authors assume no responsibility for accidents that may occur during performance of these activities.

Drs. Fifer and Ledbetter are former public school science teachers with several hundred hours of experience training elementary, middle school and high school teachers. We hope you and your students enjoy these activities. If you have any questions or comments, please contact us at Near-Normal Design and Production Studio, 1540 Keller Pkwy #108-261, Keller, TX 76248 or nearnormaldesign@gmail.com.

Using Penny Ante Science Activities

Many schools require that teachers link their instruction to some sort of learning standards. To do this, the teacher will have to provide a connection to the specific standards used by the school. The example below is based on the national standards written by practitioners from around the United States under the auspices of the National Academy of Sciences (NAS) and the National Academy of Engineering (NAE). The *Next Generation Science Standards* (2013) are made up of three Dimensions: Practices, Crosscutting Concepts, and Disciplinary Core Ideas. A Performance Expectation is the point at which these three Dimensions intersect. To access the standards go to: http://www.nap.edu/read/18290/chapter/1. You may purchase the book, download the standards or utilize it for free online. If you use an electronic version, simply choose the topic of the activity that you will teach and put that into the search function. You'll be given all of the entries for that topic; you must decide upon the objective that best fits your lesson.

Depending on how the instructor decides to use the activities, in particular the follow-up discussion of the answers to the questions associated with the activities, student learning could fall into the categories listed on this table. However, every activity could be a part of another Disciplinary Core Idea. For instance, Adaptation and Survival could be used to teach graphing, the history of science, use of camouflage, and predator/prey relationships.

Activity	Performance Expectation	Science and Engineering Practices	Disciplinary Core Ideas	Crosscutting Concepts
Clonin' Jet Liners (critical thinking)	2-PS1-1 Plan and conduct an investigation to describe and classify different kinds of materials by their observable properties.	2-PS1-2 Analyzing and interpreting data	PS1.A Structure and properties of matter	2-PS1-1 Patterns in the natural and human designed world can be observed.
High Water, Low Water (earth/space science)	MS-ESS3-4 Construct an argument supported by evidence for how increases in human population and per-capita consumption of natural resources impact Earth's systems.	MS-ESS3-4 Engaging in arguments from evidence	ESS3.C Human impacts on Earth systems	MS-ESS3-1, MS-ESS3-4 Cause and effect relationships may be used to predict phenomena in natural or designed systems.

Heart Beats (life science)	4-LS1-1 Construct an argument that plants and animals have internal and external structures that function to support survival, growth, behavior, and reproduction.	4-PS4-2, 4-LS1-2 Develop a model to describe phenomena. Use a model to test interactions concerning the functioning of a natural system.	LS1.A Structure and Function	4-LS1-1, LS1-2 Systems and System Models A system can be described in terms of its components and their interactions.
Soda Pop! (physical science)	MS-PS1-2 Analyze and interpret data on the properties of substances before and after the substances interact to determine if a chemical reaction has occurred.	MS-PS1-2 Analyze and interpret data to determine similarities and differences in findings.	PS1.B Chemical reactions	MS-PS1-4 Cause and effect relationships may be used to predict phenomena in natural or designed systems.

Table of Contents

Physical Science

Problem Solving

12

Clonin' Jet Liners

Objectives:

Students identify variables that remain constant, those being manipulated, and those that respond..

Students will build a model.

Students will make inferences and observations.

Students will collect data.

Materials:

Foam or paper airplane (available from Oriental Trading Company, 1-800-875-8480), paper, pencil, meter tape

Procedure and Results:

1. Carefully follow the directions to build your jetliner.
2. Test-fly it and make the adjustments until you think it is flying well.
3. How does the flight of your jetliner compare with that of your classmates? (Circle your choice)

    Better Than   Worse Than   About the Same As

4. What are the variables associated with the assembly of your jetliner?

5. What could be the reason(s) behind the flight differences?

6. What are ways you could control these differences?

7. Plot your test flights on the following graph. Plot Attempt versus Distance flown.

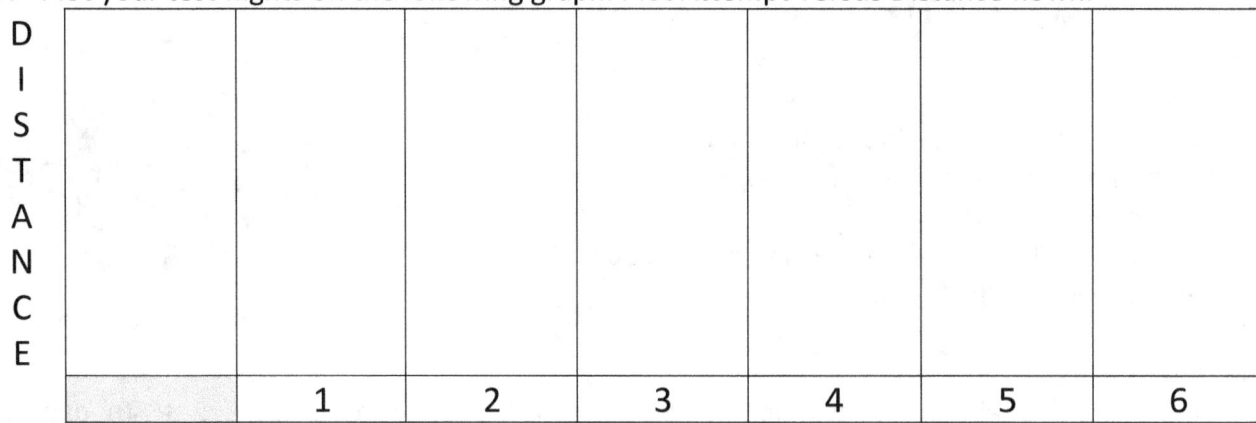

| | 1 | 2 | 3 | 4 | 5 | 6 |

ATTEMPTS

8. What is the average distance of your test flights?

9. How is this airplane like a real airplane?

10. How is it different from a real airplane?

11. Why is it important to test designs of machines?

14

Flying Fuzzies

Objectives:

Students will observe the flight patterns of different objects.

Students will predict flight patterns, given changes in variables.

Students will manipulate variables.

Materials:

Superfuzzy, paper clips, scissors, paper, markers (optional), ruler (optional).

Procedure:

1. Bend one of Superfuzzy's ears forward, and the other back, so that SF looks like a helicopter.
2. Attach a paper clip to SF's toes.
3. Hold SF above your head, drop him/her, and observe the behavior.
4. Repeat the dropping several times until you have made at least 5 different observations.
5. Using the paper and scissors, make another flying bunny, but change one characteristic (length of ears, width of body, etc.).
6. Write down how you think your change will make your bunny act differently from SF (see #3 below).
7. Repeat steps 1 through 4 with your flying bunny.

Results:

1. Make five observations about the flight behavior of SF.

2. What characteristic did you change in making your flying bunny? (Remember, you can only change ONE characteristic.)

3. How do you think this will change your bunny's flight behavior from that of SF's?

4. Make five observations about the flight behavior of your bunny.

5. What variable did you change?

6. What variable responded to that change?

7. What did you keep the same?

8. What is a variable? How do you know?

9. Just for fun, compare your flying bunny to other flying bunnies in your class. Whose flies the longest?

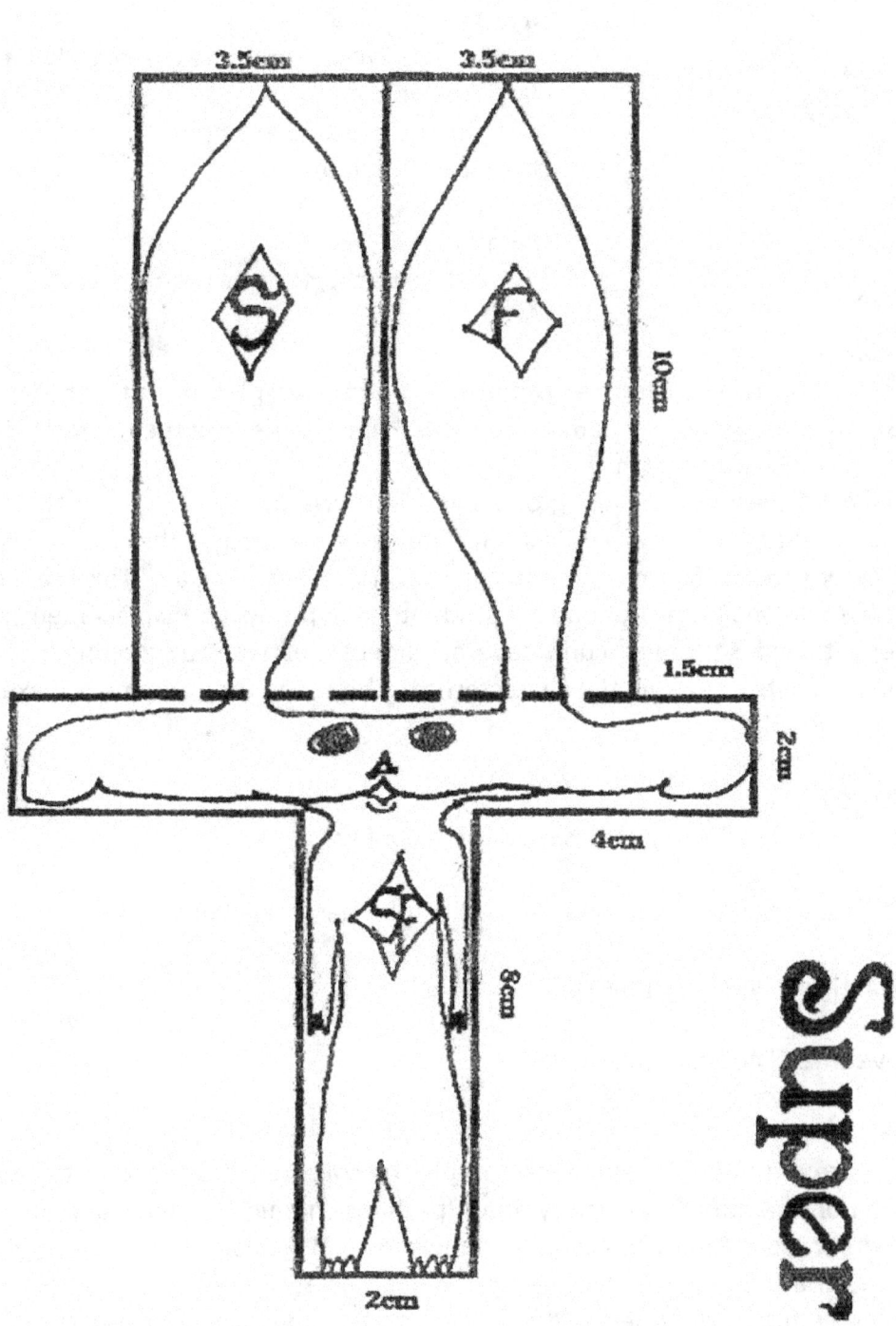

It's in the Cards
Teacher's Instructions

Objectives:

Students will observe the characteristics used in classification.

Students will give examples of items that fit into the classification scheme.

Materials:

Marker, paper, tape or thumb tacks

Procedure:
1. Before this activity prepare examples and non-examples by writing them on paper. For example, if the topic is mammals, you will have cards with the names of mammals, fish, birds, trees and so forth.
2. Make a T chart with a "yes" group and a "no" group.
3. Tape or tack the first example in the "yes" group and say, "This is a 'yes'."
4. Tape or tack the first non-example in the "no" group and say, "This is a 'no'."
5. Hold up a "yes" or a "no" and ask students to hypothesize its placement within a group.
6. Repeat steps 3 through 5 until several students seem to be grouping correctly.
7. Ask students to explain the characteristics they used for grouping the examples and the non-examples.

Results to Discuss:
1. What are the characteristics used for grouping?

2. What was the clue you used to identify this characteristic?

3. Give three "yes" examples.

4. Give three "no" examples.

Extensions:
1. Prepare the examples and non-examples by collecting magazine pictures (such as mammals and non-mammals), writing symbols (perhaps chemical symbols and formulae), or photographic slides (for example mountains and plains).
2. As a variation of this technique, use actual objects. For instance a beaker, graduated cylinder and a test tube could be "yes" examples while a cup measure, a bowl, and a water glass could be "no" examples for laboratory equipment.

Rope Trick

Objectives:

Students will hypothesize methods for removing the rope from their hands.

Students will use problem-solving skills to remove the rope.

Materials:

Two ropes, approximately 5 feet long with loops (slip knots) tied in the ends of each.

Procedure:

1. Place the loops of one rope over your partner's hands.
2. Hang the other rope over the rope your partner is now wearing.
3. Place the loops of the second rope over your hands. See illustration below.
4. Without cutting, untying, or removing the loops from your hands or your partner's hands and without untying the knots, get the ropes apart.

Results:

1. What methods did you try at the beginning of this task?

2. What method did you use to finally get the ropes apart?

3. What could you have done to escape faster?

4. How is this science?

Rope Trick
Teacher's Instructions

Objectives:
    Students will hypothesize methods for removing the rope from their hands.
    Students will use problem-solving skills to remove the rope.

Materials:
    Two ropes, approximately 5 feet long with loops (slip knots) tied in the ends of each.

The method for getting free is very simple. Take the rope that is on top and make a loop. Push this loop through the back side of the loop on your partner's opposite hand. Put your loop over your partner's hand, then stretch your hands out so that your rope pulls free. See the illustration below.

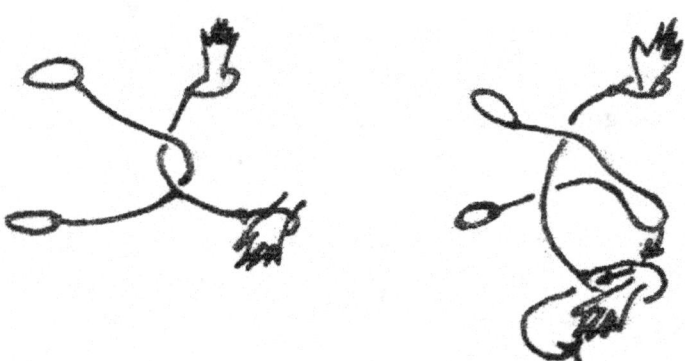

Thinking and Thinking

Objectives:
> Students will think critically about current events in science.
> Students will make decisions based on current research from scientific journals.
> Students will collect data using the Internet.

Materials:
> Access to the Internet and/or current scientific journals

Procedure:
1. Choose a topic about which you would like more information from the list below.
2. Use the Internet or scientific journals to gather information on which to base your decisions.

- Should cloning experiments be extended to humans?
- Should growth hormones be given to cattle?
- Should preservatives be added to foods with a short shelf life?
- In a recent statement, scientists commented that the space program had produced no new technology, but had expanded on already developed technology. Given this statement, should the space program funding be increased?
- Swimming with dolphins is becoming increasingly popular. Should this become a common practice?
- Many species are nearing extinction. Their only chance for survival is as zoo animals. Should they be captured and placed in zoos?
- Petting zoos are one way for children to come in contact with animals they may never otherwise see, such as sheep, goats, pigs, rabbits, and so forth. In many cases the animals are drugged to avoid possible injury to the children. Is this fair treatment of the animals?
- Researchers often use animals to test product safety. Should this use of animals be stopped or simply better monitored?

Results:
1. Once you have found your data, prepare a paper you could present to the class.
2. The paper should contain the question, your position on the answer, and the data to support your position.

Earth Science

24

Around the World in 81 Days

Objectives:

Students will determine the relationship between the densities of hot and cold air.

Students will construct a model.

Students will use logical inferences in drawing conclusions.

Materials:

7 sheets of 50 cm X 75 cm tissue paper, 35 cm of lightweight wire, charcoal (Match Light®), hot water vent pipe or two vegetable cans from the school cafeteria (gallon size), matches, scissors, glue (stick or white), clear tape, asbestos gloves, marking pens, 5 gallon bucket of water or fire extinguisher, **eye protection for each student and the teacher**

Procedure:

1. Place two sheets of tissue paper on a flat surface (one on top of the other). Fold in half lengthwise and cut along the fold. You now have 2 sheets of 35 cm X 50 cm.

2. Stack the four sheets with the 50 cm at the top. Fold from the side, making 8 sheets measuring about 35 cm X 25 cm. Fold once more, making 16 sheets about 35 cm X 12.5 cm.

3. Unfold once, making 8 sheets measuring 35 cm by 25 cm. Now fold a diagonal as illustrated. Cut along this diagonal.

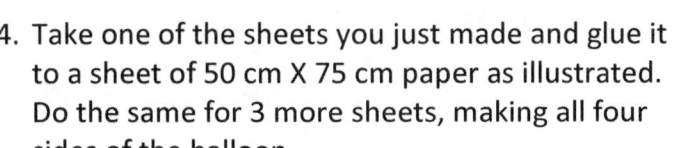

4. Take one of the sheets you just made and glue it to a sheet of 50 cm X 75 cm paper as illustrated. Do the same for 3 more sheets, making all four sides of the balloon.

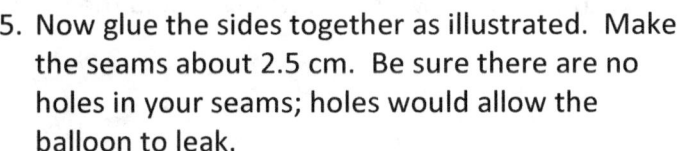

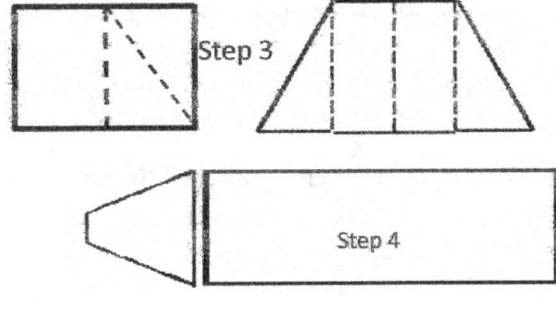

5. Now glue the sides together as illustrated. Make the seams about 2.5 cm. Be sure there are no holes in your seams; holes would allow the balloon to leak.

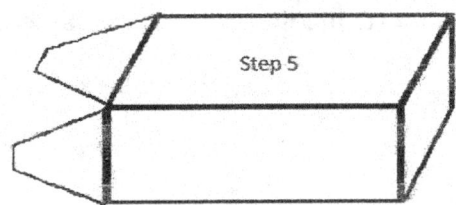

6. To make the top of the balloon, take the last sheet of tissue (50 cm X 75 cm) and cut a strip 25 cm X 50 cm (see illustration). Glue all four sides of the balloon to the top. Again, make sure you have no leaks.

7. Bend the 35 cm of wire into a loop. Place the wire loop just inside the smaller opening of the balloon and fold the tissue paper around the wire, securing it with the clear tape.

8. Decorate your balloon with markers, streamers, or whatever you wish.

9. Put on your goggles or other eye protection.

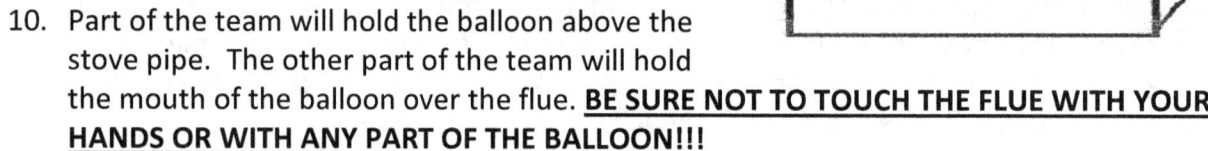

10. Part of the team will hold the balloon above the stove pipe. The other part of the team will hold the mouth of the balloon over the flue. **BE SURE NOT TO TOUCH THE FLUE WITH YOUR HANDS OR WITH ANY PART OF THE BALLOON!!!**

11. Hold the balloon in place until it fills with hot air and begins to rise.

Results:

1. As the balloon began to fill with air, what did its sides feel like? Why does it feel this way?

2. Why can't the balloon fly until you fill it with hot air?

3. Which is heavier, hot air or cold air? How do you know?

4. If a strong mass of cold air came in contact with a mass of warm air, where would the warm air go?

5. If a strong mass of warm air came in contact with a mass of cold air, where would the cold air go?

6. What happens to a balloon when the warm air inside it cools off?

7. Which way does the cold air from the arctic tend to flow?

Around the World in 81 Days

Teacher's Instructions

Objectives:

Students will determine the relationship between the densities of hot and cold air.

Students will construct a model.

Students will use logical inferences in drawing conclusions.

Materials:

7 sheets of 50 cm X 75 cm tissue paper, 35 cm of lightweight wire, charcoal (Match Light®), hot water vent pipe or two vegetable cans from the school cafeteria (gallon size), matches, scissors, glue (stick or white), clear tape, asbestos gloves, marking pens, 5 gallon bucket of water or fire extinguisher, eye protection for each student and the teacher, 13" X 8" (approximately) sheet cake pan, pliers

Procedure:

1. Bend the bottom of the pipe or one of the cans so that air can get to the charcoal. See illustration.

2. If you are using the vegetable cans, make sure both ends are cut out and that there are no rough edges. Crimp one can and slide it into the other, forming a tube. See illustration.

3. **Be sure to select a launch site clear of power lines and trees!**
4. Put the tube in the cake pan. Pour some charcoal into the tube. Use about what you would to grill a steak.
5. Put on your eye protection.
6. Light the charcoal. Wait for the flames to die down; do not wait for the coals to turn white. This takes about 10 minutes.
7. **Have a bucket of water or a fire extinguisher readily available.** At least one group of students will get their balloon too close to the heat source.
8. **Drench the charcoal if you leave it unattended.**

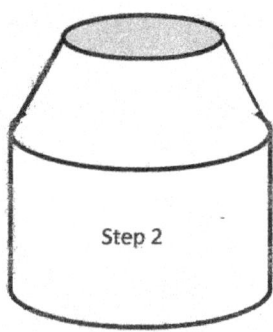

28

Davy Jones's Locker

Objectives:

Students will operationally define pressure.

Students will manipulate laboratory equipment.

Materials:

2 liter plastic soft drink bottle, plastic pipettes, hex nuts to fit pipettes, scissors, colored acetate sheets, water

Procedure:

1. Cut plastic dropper so that the tube portion measures approximately 1cm.
2. Cut a circle from the acetate sheet and punch a small hole in the center of it. Slide the circle onto the dropper. Screw the hex nut onto the dropper. (See diagram.)
3. Cut slits in the acetate circle which line up with the points on the hex nut. There should be 8 slits. Slits go from the edge of the circle to the hex nut, but no deeper.
4. Squeeze enough water into the dropper allowing its top to float just above the top of the water.
5. Fill the soft drink bottle with water. Push your diver into the soft drink bottle and screw on the top.
6. Squeeze the bottle and observe the motion of the diver.

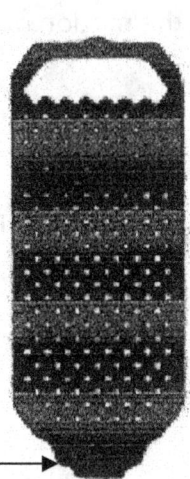

Hex Nut ⟶

Results:

1. What happens when you squeeze the bottle?

2. What happens when to the water inside the diver when you squeeze the bottle?

3. How did the acetate circle affect the diver?

4. Remove your diver from the bottle and fold the acetate circle along the slits at about a 45° angle down. (See diagram.) Put the diver back in the bottle and squeeze again.

5. Describe the motion of the diver.

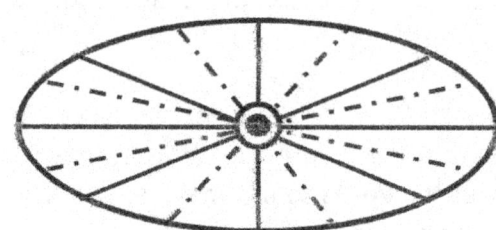

6. Remove the diver and fold the circle up at about a 45° angle. (See diagram.) Re-insert the diver and squeeze. Describe the motion of the diver.

Cut on the solid lines; fold down on the dashed lines. Be sure **NOT** to cut the lines all the way to the center. **GENTLY** poke a hole in the center.

7. Why do prop planes have propellers in front while boats have propellers in back?

30

The Devil and the Deep Blue Sea

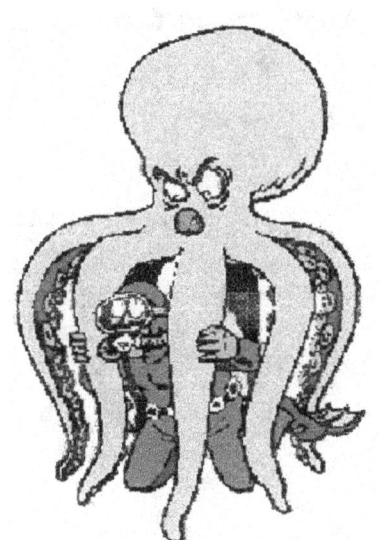

Objectives:

   Students will operationally define air pressure.
   Students will manipulate laboratory equipment.
   Students will determine what variables affect air pressure.

Materials:

   2 liter plastic soft drink bottle, plastic eye droppers, hex nuts to fit droppers, scissors, water, paper clip, mono-filament fishing line, glue gun, food coloring, heat source

Procedure:
1. Cut plastic dropper so that the tube portion measures approximately 1cm. Screw the hex nut onto the dropper. (See diagram.) Repeat this procedure for the second dropper.
2. Unbend the paper clip and make it into a hook. The hook portion must be able to fit through the mouth of the soft drink bottle. Glue the hook to one of the droppers. (See diagram.)
3. Cut off about 7.5cm of mono-filament and form it into a loop. Melt the ends of the loop together. Glue the loop to the top of the other dropper.
4. Squeeze enough colored water into the hook diver allowing its top to float above the top of the water. This diver must be able to float while holding the loop diver.
5. Squeeze enough colored water into the loop diver to keep it sitting on the bottom of the bottle. You may want to use the glue gun to seal the loop diver.
6. Fill the soft drink bottle with water. Push your divers into the soft drink bottle and screw on the top.
7. Squeeze the bottle and observe the motion of the diver.
8. Now try to hook the loop and bring both divers to the top.

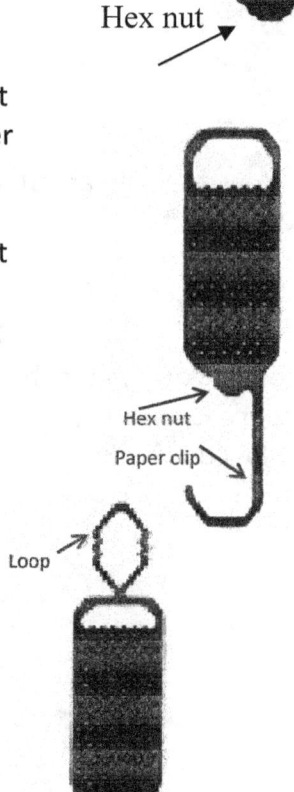

Results:
1. What makes the hook diver dive?

31

2. What happens to the water in the loop diver when you squeeze the bottle?

3. Why must you keep constant pressure on the bottle to keep the hook diver in place?

4. Why can't you squeeze the sides of the bottle until they touch?

5. How does pressure affect you daily?

6. What did you change in this experiment? What happened when you made the change?

7. How does the amount of pressure you use affect the diver?

8. Since you cannot change the amount of water in the bottle once it's sealed, what is changed by squeezing on the bottle? How do you know?

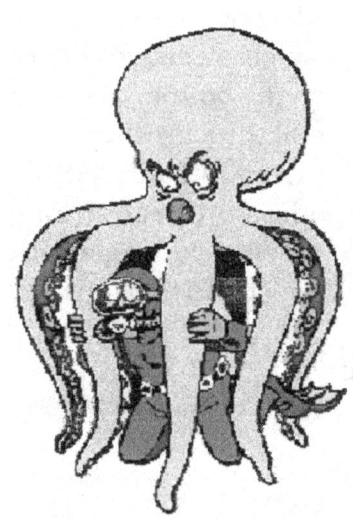

High Water, Low Water

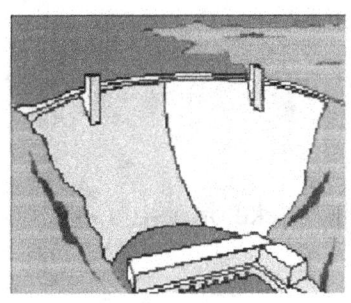

Objectives:

Students will operationally define flow rate.

Students will construct a model of a paddle wheel.

Students will relate water flow to energy produced.

Students will determine the relationship between use of water and environmental impact.

Materials:

1 wire coat hanger, 1 wire cutter, 1 meter of rubber tubing, 1 transparent bottle (two-liter cola bottle), 1 shallow pan or dish, 1 piece of Styrofoam® (20cm X 20cm X5cm), 1 plastic knife (serrated edge), silicone sealer, ice pick or awl, pad of cardboard

Procedure:
1. Use the plastic knife to cut the Styrofoam into a paddle wheel shape similar to the illustration.
2. Punch a hole through center of the paddle wheel. **Caution: always push the ice pick or awl away from hands or other body parts and into the pad of cardboard.**
3. Using the wire cutters, make an "axle" for the paddle wheel with the coat hanger. It must be able to go across your pan or dish.
4. Insert wire "axle" through the paddle wheel and test for free spinning.
5. Punch a hole in the bottle cap large enough to put the rubber tubing through. **Caution: always push the ice pick or awl away from hands or other body parts and into the pad of cardboard.** Attach rubber tubing to the bottle top using silicone sealer if necessary to secure the tubing and prevent leaks.

Water Wheel

6. Fill bottle with water. Invert bottle and punch an air hole in bottom. **Caution: always push the ice pick or awl away from hands or other body parts and into the pad of cardboard.** By placing your fingertip over this hole and releasing it, you will be able to control most of the flow of the water from the bottle down through the tubing.
7. Hold tubing so that when the water is released it will flow onto the paddle wheel.

Results:
1. How does the height of the water affect the paddle wheel?

2. How does the amount of water affect the speed of the paddle wheel?

3. How do these observations (in 1 and 2) relate to the water level in a lake or reservoir?

4. What types of flow rates could you develop from the height of the water source versus the rate of revolutions of the paddle wheel?

5. How is your model like the generators used to produce electricity?

6. If the turning of a wheel is used to generate electricity, how does this impact the environment in and around a lake?

7. Where do dams fit into the generation of electricity? How does this impact the environment?

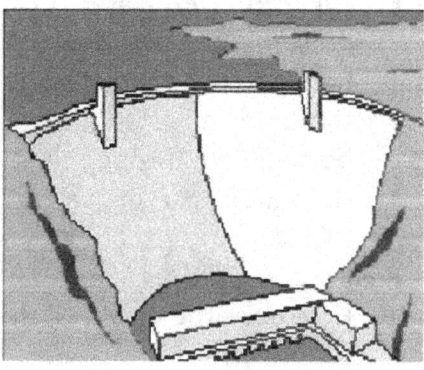

How Far Can You Throw a Piano?

Objectives:

Students will observe differences in soil grain size.
Students will predict the carrying capacity of the water.

Materials:

Aluminum loaf pans with a small hole in one end, plastic spoons, hand lenses, permanent markers, 100 ml beaker with water, sand, bucket to catch runoff, newspaper, book

Procedure:

1. Fill the loaf pan about 6 cm deep with damp sand.
2. Form the sand into a wedge shape in the end of the pan without the hole. The wedge should be deepest at the end of the pan and thin out to the center of the pan.
3. Set the pan on the end of a table with the hole hanging over the edge. Place the bucket on newspaper under the hole. When you add water, it will run slowly out of the hole and into the bucket.
4. Elevate the end of the pan without the hole by placing it on a book.
5. Draw a circle about half the size of a dime on the inside of the spoon.
6. Pour not more than 100 ml of water, 10 ml at a time onto the thickest part of the sand.
7. Pick up some of the material from the center part of the stream with the spoon. You only need as much as covers the circle in your spoon.
8. Examine it with a hand lens. Draw what you see in the space below.
9. Rinse off the spoon.
10. Repeat steps 6 and 7, but get the sand from the edge of the stream.

Results:

1. Draw what you see in the circles below.

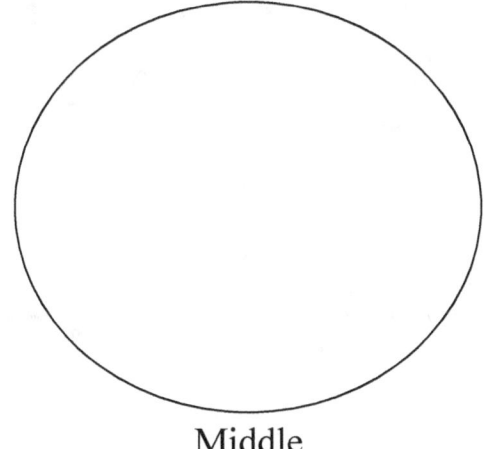

Middle

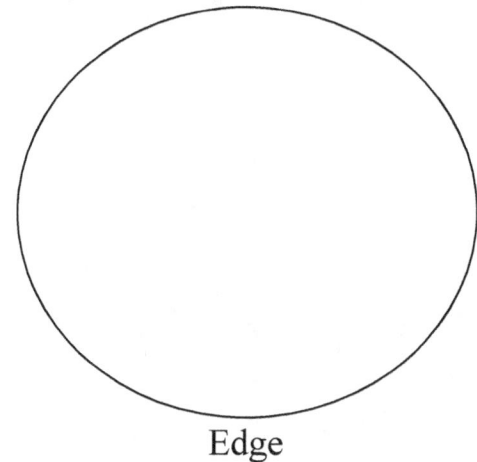

Edge

35

2. How are these samples the same?

3. How are they different?

4. If the smaller particles came from the center of the stream, and the larger ones came from the edge, what can you say about the water's ability to lift weight?

5. If you fell out of a boat into a river, which part of the stream would push you harder, the edge or the center? Why?

How Far Can You Throw a Piano?
Teacher's Instructions

Objectives:
   Students will observe differences in soil grain size.
   Students will predict the carrying capacity of the water.

Materials:
   Aluminum loaf pans with a small hole in one end, plastic spoons, hand lenses, permanent markers, 100 ml beaker with water, sand, bucket to catch runoff, newspaper, book

Procedure:
1. Use a pencil, pen or other somewhat sharp object to poke a hole in one end of the loaf pan. These are the inexpensive pans made from aluminum foil.
2. Choose a sand with varying particle sizes. Play sand works well. The sand should be damp enough to stick together, but not wet.
3. The sand the students want to examine is at the end of the out-wash area, rather than in the middle of the wedge.
4. You may prefer to use one stream table (aluminum pan) rather than allow each group of students to make their own. The amount of sand they will need is small and should be obtainable from one set-up.

How Hard Is It?

Objectives:

    Students will measure amounts of water.

    Students will differentiate between "hard" and "soft" water.

    Students will infer the sources of materials in water.

Materials:

    3 water sources, dishwasher soap, graduated cylinders, 4 beakers, water droppers, balance, 4 plastic spoons

Procedure:

1. Collect 30 ml of water from each water source and put these in labeled beakers. Observe these samples carefully. Report your observations in the Results section.
2. In the fourth beaker, mix 1 gram of soap powder with just enough water to dissolve it.
3. Using the dropper, add one drop of soap solution to each beaker. Stir vigorously for 30 seconds. Keep a record of the number of drops of soap solution used in each beaker in the chart below. Note: If all samples bubble for more than 30 seconds, repeat the experiment using a toothpick rather than a dropper to add a smaller amount of soapy water.
4. Repeat step 3 until soap bubbles remain for 30 seconds after you quit stirring. Do not add any more soap to the beakers!

Results:

1. What is the source of each of your water samples?

2. How does each sample appear?

3. How does each sample smell?

Sources			
Number of Drops			

4. Which water had bubbles for 30 seconds first?

5. To which water did you add the most soap?

6. Soap does not make suds as easily in "hard" water as it does in "soft". Which water do you think is the "hardest"? Why do you think that?

7. What do you think makes water "hard"? Why do you think this?

Ka-Boom!

Objectives:
    Students will state an operational definition of air pressure.
    Students will practice laboratory safety.
    Students will interpret data.

Materials:
    Empty soft drink can, pan of water, beaker tongs, heat source, eye protection, metric ruler

Procedure:
1. Measure the diameter and height of the soft drink can. Record these measurements in the Results section.
2. Pour a small amount of water into the can. Pour enough water into the pan to cover 2.5 cm of the can when it is placed vertically in the water.
3. Heat the can until all the steam disappears.
4. Grasp the can with the beaker tongs and quickly invert it into the pan of water.
5. Move away from the pan.

Results:

Can Diameter	
Can Height	
Can Radius	

1. The surface area of the end of the can is the radius multiplied by itself, then multiplied by 3.14.
   The surface area of one end of the can is:

2. Multiply the answer in question 2 by two to find the surface area of both ends of the can. The total end surface area is:

3. The surface area of the side of the can is the diameter multiplied by 3.14, then multiplied by the height of the can. The surface area of the side of the can is:

4. To find the total surface area of the can, add the numbers you got in questions 3 and 4 together. The total surface area of the can is:

5. Describe the action of the can when you inverted it into the water.

6. What caused the can to crush?

7. What happened to the air inside the can?

8. If air pressure at sea level is 2.7 kg/cm^2, how many kg did it take to crush the can? (Hint: How many square centimeters are there in the surface area of the can?)

9. Is there air pressure inside your body? How do you know?

10. What would happen if an empty sealed soda can were heated? How do you know?

11. What does air pressure have to do with this activity?

12. What is air pressure?

Powerful Gases

Objectives:

    Students will predict the effect of keeping gases in sealed containers.

    Students will operationally define effervescence.

Materials:

    Baby bottle, small balloon, soft drink

Procedure:

1. Try to blow up the balloon in one breath, without stretching the balloon beforehand.
2. Cut a hole in the plastic plug of the baby bottle of sufficient size to allow the balloon to fit snugly. Be sure the lip of the balloon cannot pass through the hole.
3. Fill the bottle half full of the soft drink, attach the balloon and close tightly.
4. Shake the bottle.

Results:

1. Could you blow up the balloon easily?

2. What made the balloon expand?

3. What is the gas in soft drinks?

4. Why does the gas bubble out of the liquid?

5. Does gas have mass? How do you know?

A Recipe for Boiling Water

Objectives:

Students will hypothesize reasons for water boiling at a lower temperature.

Students will make observations.

Materials:

Flask, stopper, hot plate, aquarium, pot holder

Procedure:

1. Fill the flask 1/3 full of water.
2. Bring the water to a boil. **Caution: The flask will become too hot to handle safely.**
3. Using the pot holder, remove the flask from the heat and quickly put in the stopper.
4. Allow the flask to stand for a few minutes, then dip it in the aquarium water.

Results:

1. What happens when the flask is dipped in the aquarium?

2. Why does this happen?

3. What happens to the air pressure inside the flask when the water is boiling? How do you know?

4. Would you get the same results if there were no stopper in the flask?

5. Why does it take longer to boil eggs at a higher altitude?

Sea Hunt

Objectives:

Students will acquire data through their senses.

Students will use logical inferences in drawing conclusions.

Students will operationally define density.

Materials:

25 ml plastic pipettes, 2 L clear plastic soda bottles, 10/32 hex nuts, scissors

Procedure:

1. Cut the tip off the pipette about 1 cm below the bulb.
2. Carefully screw the hex nut onto the remainder of the pipette.
3. Now you have your "deep sea diver". By squeezing the bulb with the tip immersed in water, you can make the diver as heavy or light as needed.
4. Fill the soda bottle with water until nearly full. Insert the diver and fill completely full, trying to remove all the air possible. The diver should float at the top of the bottle. Replace the cap on the bottle.
5. Practice squeezing the bottle. You should be able to move the diver up and down, depending on the amount of pressure you exert on the sides of the bottle when you squeeze.

Hex Nut

Results:

1. What happens to the volume of water in the bottle when you squeeze?

2. If the mass of the water is unchanged and you diminish the volume of water, what happens to the density? (Hint: Density = Mass/Volume)

3. What would happen to a ship if it sailed into water that was less dense than that it had been sailing in?

4. What would happen if the water were denser?

5. Practice making the diver hover at some point in the water. What has happened to allow the diver to stay in this position?

6. In what situations would it be important to know the density of a liquid?

Solid Bubbles Demonstration

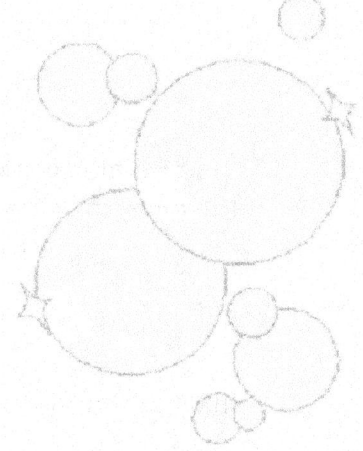

Objectives:
Students will observe apparent changes in optical density.
Students will compare the bubble film to the atmosphere.

Materials:
Commercially prepared bubble liquid, bubble wand, overhead or other type of light projector, thermometer, sunny area

Teacher's Procedure:
1. Dip the bubble wand in the solution. Hold this in front of the upper light on the overhead projector (or other projector light).
2. Project the image.

Results 1:
1. Are there parts of the film through which light cannot pass? What causes this?

2. If you look directly at the film, can you see through it? Why?

Students' Procedure:
1. Find a sunny spot either in the room or outside.
2. Measure the air temperature in the direct sunlight and in an area of indirect sunlight. Note these temperatures in the Results 2 section.

Results 2:
1. What is the temperature in the direct sunlight?

2. What is the temperature in the indirect sunlight?

3. Why are the temperatures different?

4. If the film of bubble solution were a model of the atmosphere, at which spot would you expect the temperature to be higher, at the edge of the bubble wand or at the center? Why?

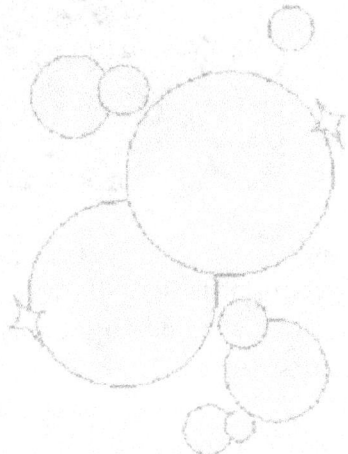

Stream Speed

Objectives:
    Students will measure the speed of flowing water using a stopwatch and a meter stick.
    Students will predict changes in flow rates based on the amount of water in the stream.

Materials:
    Flowing water (at least 10 meters long), 10-meter tape or 10 m of string, stopwatch, small stick or float

Procedures:
1. **Review safety procedures with students before going to the stream and remind them once they get to the test area.**
2. Pick a beginning spot on the stream bank. Station the student with the stick and another student at this spot.
3. Measure downstream 10 meters. Station the student with the stopwatch here.
4. Have one student yell, "GO!" to signal the other students to drop the stick and start the watch.
   As the stick passes, the student stops the watch. Another student should record the time.
5. Repeat this process 4 more times. If you use different sticks each time, make sure that they are about the same size and shape. Ignore trials in which the stick hangs on a rock or other debris.

Results:

Trials	Seconds
1	
2	
3	
4	
5	
Total	
Average	

1. Divide 10 meters by the AVERAGE to get meters/second.

2. Are there parts of the stream which flow faster than others do? Describe these areas.

3. Describe the slower flowing areas of the stream.

4. What would happen to the rate of flow if there were more water in the stream? How do you know?

5. What would happen to the rate of flow if there were less water in the stream? Why?

6. Why is it important to know how fast a stream flows?

Temperatures--High and Low

Objectives:

Students will use thermometers to determine temperatures of different parts of the environment.

Students will hypothesize reasons for temperature variation.

Materials:

Three thermometers, meter stick or 1 meter of string, stopwatch or clock with a second hand, small shovel or trowel.

Procedures:
1. Measure a meter above the ground, up the side of a tree for instance. Make sure this is not in direct sun. Hold the thermometer in place for approximately 5 minutes. Record the temperature.
2. Using the trowel, remove part of the leaf layer and lay the thermometer on the ground. Cover it with the leaves. After 5 minutes record the temperature.
3. Lay the third thermometer on top of the ground out of direct sun. In 5 minutes, record the temperature.

Results:

Location of Thermometer	Temperature in C°
Air	
Leaf Layer	
Ground	

1. Why do you think some animals live under the leaf layer?

2. Where would you go if you wanted to be warmer in the winter?

3. Would the temperature under the leaf layer be warmer or cooler than the air temperature in the summer? How do you know?

4. Why do you think there is a temperature variation among the places you measured?

5. What does this activity have to do with energy use for heating and cooling? Why?

6. How does energy conservation affect animals in the environment?

7. How does energy conservation affect you?

Up, Up, and Away

Objectives:
   Students will observe air currents.
   Students will operationally define density.

Materials:
   Commercially prepared bubble liquid, bubble wand, compass.

Procedure:
1. Go outside or to an area where you feel air movement.
2. Blow bubbles and observe their direction of flight. Repeat this step several times.

Results:
1. Which way did the bubbles go first?

2. Did the bubbles change directions as they rose?

3. Which way was the wind blowing?

4. Were there any man-made objects in the area? How did they affect the flight of the bubbles?

5. If no wind was blowing what would the bubbles do?

6. Why do the bubbles seem to float on the air?

Which Way Do I Go?

Objectives:
    Students will predict results from data given.
    Students will manipulate lab materials.

Materials:

Pizza cardboard, 10 5 cm X 25 cm corrugated cardboard strips, coat hanger, marking pen, protractor, scissors, pliers, masking tape, upper air wind conditions from the local airport or from NOAA web site (http://www.nhc.noaa.gov/index_station.shtml)

Procedure:
1. Using your protractor and marking pen, divide the pizza cardboard into compass points: N-360, S-180, E-90, W-270 and label them. You may wish to subdivide the rest of the cardboard into increments of 10 degrees.
2. Take the 10 cardboard strips and cut arrow points on one end of each. Label the strips 3000, 6000, 9000, 12000, 15000, 18000, 21000, 24000, 27000, 30000. Mark the center of each strip.
3. Bend the coat hanger in the shape illustrated.
4. Insert the coat hanger through the center of the pizza cardboard. Secure it to the bottom with tape.
5. Push the coat hanger through the center corrugation of each cardboard strip.
6. Now you have ten arrows that can be used to show wind direction in the upper altitudes from 3000 to 30000 feet.
7. Examine the weather report supplied by your teacher regarding the upper altitude winds and directions. Using your pizza cardboard compass, arrange your arrows from the wind directions for each altitude.

Results:
1. From which direction is the wind blowing at each of the given altitudes?

Altitude	Direction	Altitude	Direction	Altitude	Direction
3,000		6,000		9,000	
12,000		15,000		18,000	
21,000		24,000		27,000	
30,000					

2. You live in Dallas and want to fly to San Antonio. If San Antonio is located at 190° on the compass, what would be the best altitude to fly today?

3. What would be the best return path?

4. What would you need to determine the best altitude for flying tomorrow? Why?

5. Why is it important to know which way the wind is blowing for people other than airplane pilots?

Life Science

The Bean Stalk

Objectives:

Students will measure and graph the leaves of bean plants.

Students will determine the average size of plant leaves from their data.

Students will collect and interpret data.

Materials:

5 to 10 healthy bean plants, metric rulers

Procedures:

1. Each student will measure 5 leaves from each plant and record those measurements. Round off to the nearest centimeter.
2. Plot the measurements of each leaf on the bar graph below.
3. Enter your data on a class frequency table before you answer the questions in the Results section.

Results:

Length in cm
↓

15					
14					
13					
12					
11					
10					
9					
8					
7					
6					
5					
4					
3					
2					
1					
0	1	2	3	4	5

Leaves

1. How long was the longest leaf measured?

2. How long was the shortest leaf measured?

3. What was the most common leaf length?

4. If you walked out into a field of bean plants, what do you think the most common leaf length would be? Why do you think this?

The Bean Stalk
Teacher's Instructions

Objectives:
Students will measure and graph the leaves of bean plants.

Students will determine the average size of plant leaves from their data.

Students will collect and interpret data.

Materials:
5 to 10 healthy bean plants, metric rulers.

Procedures:
1. Obtain bush beans. You could have the students raise the beans, keeping a record of leaf growth and stem height.
2. If you are just measuring the leaves, plants should be at least 10 cm tall and you may want to direct the students to only measure leaves that are one or more centimeters in length.
3. Use the frequency chart (or T-chart) to answer the questions at the end of the activity.

Length of leaves (cm)	Number of leaves
15	
14	
13	
12	
11	
10	
9	
8	
7	
6	
5	
4	
3	
2	
1	

Big or Little? Tall or Small?

Objectives:

    Students will devise a classification system.

    Students will give reasons for their methods of dividing the beans/peas.

Materials:

    Egg carton, selection of dried beans/peas, small sticky notes on which to write Group Numbers

Procedure and Results:

1. Pour your collection of beans/peas into the lid of your egg carton.
2. Examine your beans/peas and divide into two groups, using the other divisions in the egg carton. Use the sticky notes to label them Group I and Group II.
3. How could you name Group I?

4. How can you distinguish Group I from Group II?

5. How could you sub-divide Group I? Label these Group IA and Group IB.

6. How could you sub-divide Group II? Label these Group IIA and Group IIB.

7. Could you think of another way to divide Groups I and II? How? Move the A and B labels to these new groups.

8. How does your selection of divisions and subdivisions relate to other classification systems?

9. Sub-divide your Group IIA and your Group IIB, and then sub-divide those groups again. What are the characteristics of these new sub- and sub-sub divisions?

10. Will your new division system allow you to find a single bean/pea? Why or why not?

11. How are classifications helpful to scientists?

12. Give an example of how you use classifications in your daily life.

61

The Blind Spot

Objectives:
    Students will indirectly observe the connection of the optic nerve to the eye.
    Students will make inferences regarding the placement of the optic nerves in their eyes.

Materials:
    3" x 5" index card with a 1" x 1" square drawn in the center of the left half of the card and a 1" diameter dot drawn in the center of the right side of the card

Procedure:
1. Cover your right eye with your hand.
2. Hold the card at arm's length.
3. Focus on the shape on the right side of the card.
4. Slowly bring the card toward your face until the shape on the left side of the card disappears.
5. Follow steps 1 through 4 for your left eye, focusing on the left side of the card.
6. Follow steps 1 through 4 with both eyes open.

Results
1. For your left eye, at about what distance did the shape disappear?

2. For your right eye, at about what distance did the shape disappear?

3. Was the distance the same for both eyes?

4. There are no rods and cones covering the optic nerve. What does the disappearing shape tell you about the placement of rods and cones in your eye?

5. What does the disappearing shape tell you about the optic nerve in your eye?

6. How does your body compensate for your blind spot?

7. If you had only one eye, how would your blind spot affect your daily life?

63

The Eyes Have It

Objectives:

   Students will compare monocular and binocular vision.

   Students will determine which of their eyes is dominant.

Materials:

   2 paper towel tubes, 2 different pages of a book

Procedures:

1.   Place the book on the desk in front of you.
2.   Hold the tubes near your eyes, looking through them as if they were binoculars.
3.   Focus on different pages of the book. Attempt to read the print.

Results:

1.   Which page was easier to read?

2.   Did you have to shut one eye to read the more difficult page?

3.   Why is it difficult to read both pages at the same time?

4.   From this experiment, which eye do you think is dominant (the one that does the most work)?

5.   Since one eye is dominant, why do we need both eyes?

Fastest Gun in the West!

Objectives:

    Students will measure their reaction time.

    Students will collect data.

    Students will graph data and interpret the graph.

Materials:

    Reaction time stick

Procedure:

1. Practice placing your hand at the starting line and having your partner release the stick.
2. By reading the value on the reaction time chart, you can see how fast you reacted.

Results:

1. What was your reading on the first trial? Second trial?

2. Do ten trials and record them on chart below.

Trial	Time (sec)	Trial	Time (sec)
1		6	
2		7	
3		8	
4		9	
5		10	

3. What is your average time?

4. Who is the fastest in your group? What is his/her time?

5. Who is the slowest? What is his/her time?

6. Did reaction times improve as your number of trials increased? Why do you think this happened?

7. Write your average time on the class chart provided by your teacher. From that chart, plot a bar graph of the data from the class.

Students' Initials ↓							
Reaction Times →							

8. What is the most common reaction time in the class?

9. Is there any relationship between reaction time and athletic ability? Explain.

10. Is there any relationship between reaction time and height? Explain.

11. Is there anything you can do to improve your reaction time? How do you know?

Fastest Gun in the West! Teacher's Instructions

Objectives:
Students will measure their reaction time.
Students will collect data.
Students will graph data and interpret the graph.

Materials:
Meter stick (or yardstick), reaction time chart, scissors, glue, hand saw

Procedure:
1. Cut the strips along the solid, lengthwise lines.
2. Glue the first strip with A at the bottom of a meter stick.
3. Glue the rest of the strips in order from the smallest number in the parentheses to the largest number in the parentheses.
4. Cut the meter sticks just longer than the reaction time strips.
5. The times were calculated using the following formula:
   $s = \frac{1}{2} GT^2$
   s = distance
   G = gravity (32 ft./sec^2) or (9.8m/sec^2)
   T = time
6. The numbers in parentheses are distances. The decimals are the reaction times.
7. Prepare a class data table, either on a transparency sheet or on the chalkboard. The chart should include students' initials and average reaction times.

.125	.250	.331
(02)	(11)	(20)
.102	.239	.323
(01)	(10)	(19)
.072	.228	.315
START HERE	(09)	(18)
	.217	.306
(08)	(17)	
	.204	.298
(07)	(16)	
	.191	.289
(06)	(15)	
	.177	.279
(05)	(14)	
	.161	.270
(04)	(13)	
	.144	.260
A	(03)	(12)

Heart Beats

Objectives:

Students will count their heart beats for 10 seconds.

Students will count an animal's heart beat for 10 seconds.

Students will hypothesize reasons for differences in heart rates.

Students will predict different heart rates for different size animals.

Materials:

Mammals, lizards, birds or other classroom animals that are safe for students to hold, clock with a second hand or stop watch

Procedure:

1. Find your pulse in your wrist or neck.
2. When your teacher says, "GO", begin counting the beats until signaled to stop. Write down your number.
3. Do this two more times and record those numbers.
4. **Follow your teacher's directions for working with the classroom animals. Handle all animals gently and with respect! Do not squeeze the animal.**
5. Using the tips of your fingers, gently touch the animal's chest. You might rather let the animal's chest rest in the palm of your hand. In either case, you should feel its heart beating.
6. Repeat steps 2 and 3.

Results:

	Human	Animal
Trial 1		
Trial 2		
Trial 3		
TOTAL		
Average		
Beats/Minute		

1. Divide the human total by 3 to get the human average.
2. Divide the animal total by 3 to get the animal average
3. Multiply the human average by 6 to get beats/minute.

4. Multiply the animal average by 6 to get beats/minute.
5. Which is bigger, you or the animal?

6. Which has the faster heartbeat, you or the animal?

7. What is the relationship between the size of an animal and the speed at which its heart beats?

8. What if the animal were as large as you? How would this affect its number of heart beats in a minute?

9. Does an elephant have a faster or slower heart rate than you do? Why do you think so?

10. Does a hummingbird have a faster or slower heart rate than you do? How do you know?

11. Why do you think heart beat rate is related to size?

70

Heavy Duty

Objectives:
    Students will measure weights using a scale.
    Students will predict the weight of food needed for each animal.

Materials:
    Fish*, mammals, reptiles, birds, food for each animal, pan or beam balance, shallow container, cotton, water

Procedure:

1. **Follow your teacher's directions for working with the classroom animals. Handle all animals gently and respectfully! Do not squeeze the animal.**
2. Weigh each animal; record the weight to the nearest gram.
3. Find out how much food each animal should have per day, then weigh that food. You could look at the label on the food, ask a veterinarian, or look on the internet to find how much the animal should eat. Be sure to record the weight of the food.
4. Take the weight of each animal and divide it **into** the weight of the food it eats. Multiply this number by 100. This tells you the percentage of its body weight it eats each time it's fed. Record this on your chart.

Results:

Animal	Weight	Food Weight	Percentage

1. Which animals needed less food than you thought they would?

2. Which animal needs the most food?

3. Which animal weighs the most?

4. Why do you think some animals need more food than others?

5. If you feed the heaviest animal every day for a year, how much will it have eaten by the end of the year?

6. How much does the food you eat in one day weigh? In a year?

7. What is the relationship between the size of the animal and the percentage of food to body weight it must eat?

*To weigh fish, moisten enough cotton to cover the fish's gills and head. Weigh the cotton and a shallow container. Put the fish in the container, being sure to cover its head and gills with the moist cotton. **Once you have determined its weight, return it to the fish tank immediately.** Subtract the weight of the cotton and container from the total weight to find out how much the fish weighs.

How Does Your Garden Grow?

Objectives:

Students will design a garden.

Students will predict which plants should be planted next to each other.

Materials:

Various flower and/or vegetable seeds, paper, pencil, shovels, hoes, other gardening implements as necessary

Procedure:

1. Choose at least 4 different kinds of seeds. From the back of the seed packet, note the height, growing seasons, growth rate, and other information you think might be important (such as the amount of water or light).
2. From the information in your chart, diagram the placement of your seeds.
3. After your teacher approves your diagram, plant your seeds.

Results:

Plant	Height	Seasons	Rate	Other

1. Which plants should come up first?

2. Which plants should be the tallest?

3. Which plants should be watered the most?

4. Which plants need the most light?

5. How soon will you know if your answers are right?

6. Why are light, water, and placement important?

Left Eye, Right Eye

Objectives:

Students will determine which is his/her dominant eye.

Students will determine the relationship between dominant eye and hand-eye coordination.

Materials:

3" x 5" index card with a 1" x 1" square drawn on the right half of the card and a 1" x 1" x 1" triangle drawn on the left half of the card

Procedure:
1. Hold the card in both hands at arm's length.
2. Stare at the square and the triangle.
3. Slowly bring the card close to your face without turning your head or moving your arms to the right or left.
4. At some point you will be able to focus on only one symbol; record your observation.

Results:
1. Which symbol did you keep in sight the longest?

2. Which eye is your dominant eye? How do you know?

3. If you are right eye dominant and left handed, how could this affect your hand-eye coordination?

4. If you are right eye dominant and left handed, how could this affect your reading ability?

5. Think about questions 3 and 4. What must the brain do to control your vision and your ability to perform tasks?

75

Link Ups

Objectives:

Students will identify elements of a food chain.

Students will explain man's impact on the environment.

Materials:

Yarn, plastic bags, popcorn

Procedure:

1. Each student will have a piece of yarn tied around his/her waist and **visible** to the other students. Yellow are grasshoppers, green are guinea fowl, and brown are bobcats.
2. The popcorn represents plants and can only be eaten by the grasshoppers. The grasshoppers may only be eaten by the guineas, and the guineas may only be eaten by the bobcats. The plastic bags are the stomachs of the grasshoppers; they put the popcorn into their stomachs. **Do not actually eat any of the popcorn**.
3. If a guinea eats a grasshopper, the grasshopper must give his/her stomach to the guinea. The same thing happens when a bobcat eats a guinea. As soon as an animal is eaten he/she must go to a designated spot and sit down. **This is not a game of tag; no running, tripping or grabbing.**
4. At the end of 2 minutes (or earlier if only the bobcats are left) have all the students with a stomach come to the teacher.
5. Keep a record of the number of plain and colored pieces of popcorn.

Results:

Rounds	Grasshoppers		Guineas		Bobcats	
	Plain	Colored	Plain	Colored	Plain	Colored
1						
2						
3						
4						
5						

Results:

1. What are the different levels in the food chain?

2. What is the lowest level of the food chain?

3. What animal cannot be eaten?

4. What would happen if there were no guineas?

5. The colored popcorn represents an insecticide. If a grasshopper eats one colored piece, it will die. If a guinea eats a mixture of half colored and half plain pieces, it will die. If a bobcat eats more than half colored pieces, it will grow weak and may die due to lack of ability to hunt. From the data, which animals feel the effects of man's use of insecticides?

6. What are some alternatives to insecticides?

7. What do most food chains start with?

8. Is man part of a food chain? How do you know?

77

Link Ups
Teacher's Instructions

Objectives:
   Students will identify elements of a food chain.
   Students will explain man's impact on the environment.

Materials:
   19 meters yellow yarn, 6 meters green yarn, 2 meters
brown yarn, sandwich bags, 6 L plain popped corn, 1 L colored popped corn, timer, chart paper
for data collection (optional), marker (optional)

Procedure:
1. Make 1 tie for each student in the class; each will be from ¾ to 1 meter long. They are to be tied around the children's waists, and **visible**. Make about 16 with yellow (grasshoppers), 6 with green (guinea fowl), and 2 with brown (bobcats), depending on the class size.
2. The popcorn represents plants and can only be eaten by the grasshoppers. The grasshoppers may only be eaten by the guineas, and the guineas may only be eaten by the bobcats. The plastic bags are the stomachs of the grasshoppers; they put the popcorn into their stomachs. **Remind the students not to actually eat any of the popcorn**.
3. If a guinea eats a grasshopper, the grasshopper must give his/her stomach to the guinea. The same thing happens when a bobcat eats a guinea. As soon as an animal is eaten he/she must go to a designated spot and sit down. **Remind the students that this is not a game of tag and of the safety rules such as no running, tripping, grabbing and so forth.**
4. Scatter the popcorn on the ground. Set the timer for two minutes.
5. Give the grasshoppers 30 seconds of free hunting time, then release the guineas. Release the bobcats 20 seconds later.
6. At the end of 2 minutes (or earlier if only the bobcats are left) have all the students with a stomach come to the teacher. For the food chain to be balanced there should be one of every animal left.
7. Examine the stomachs. Each must have at least one piece of popcorn in it or the animal starved to death and was <u>not</u> eaten by the predator.
8. Have children count the number of colored popcorn kernels and the number of plain kernels. **Do not tell the students why the popcorn kernels are different colors**.
9. Keep a record of the number of plain and colored pieces of popcorn.
10. Play several more rounds, allowing children to have different roles.

Out with the Old; In with the New
Teacher's Instructions

Objectives:
Students will identify the major parts of the circulatory system.
Students will list materials carried by the blood.
Students will make a model of the circulatory system.

Materials:
Sandwich signs labeled heart, lungs, blood, oxygen, food, carbon dioxide, cells (5); masking tape, large garbage bag stuffed with wadded up newspaper and labeled waste

Procedure:
1. Choose 11 students and hang the signs around their necks.
2. Mark off the floor and arrange the children as shown. Note: you do not need to mark the arrows, as long as you keep the blood moving the indicated direction.
3. Help the students act out the process of circulation.
   - Blood and CO_2 hold hands and stand by the Heart. They then walk to one Lung.
   - Blood gets O_2 and leaves CO_2. CO_2 may go sit in the audience to reinforce the idea that it has left the body.
   - Blood and O_2, holding hands, walk down the blood vessel toward the cells. Along the way Blood gets Food and the three components proceed to the Cells.
   - Blood leaves O_2 and Food with the Cells and picks up Waste, then heads back toward the Heart.
   - On the way, Blood drops off Waste **outside** the masking tape, showing that it has left the body.
   - Blood returns to the Lung, giving it food and O_2, and then to the Heart.
4. As a variation, you may give slips of paper to the Cells, Lungs, and Heart labeled, "CO_2"; the Blood carries slips that say, "Food". The Blood exchanges papers with the Cells, Lungs and Heart.

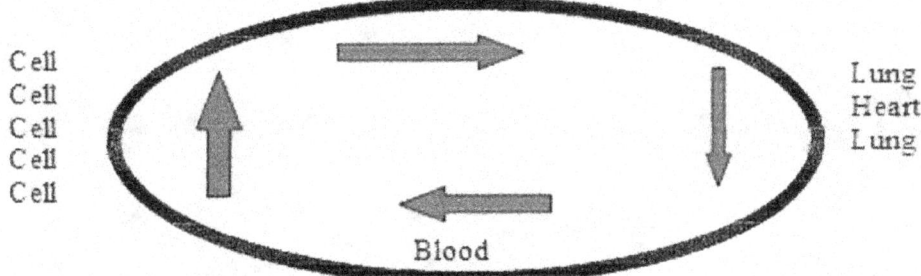

79

Out with the Old, In with the New

Objectives:

 Students will identify the major parts of the circulatory system.

 Students will list materials carried by the blood.

 Students will make a model of the circulatory system.

Materials:

 Sandwich signs labeled heart, lungs, blood, oxygen, food, carbon dioxide, cells (5); masking tape, large garbage bag stuffed with wadded up newspaper and labeled waste.

Procedure:

1. Your teacher will assign you a role to play. Follow directions carefully.
2. Make sure you observe how each of the other actors performs his/her role.

Results:

1. What does the blood carry?

2. What do the cells want to get rid of?

3. Is CO_2 a waste? How do you know?

4. Would the blood flow if there were no heart? How do you know?

5. What do cells need?

6. Is blood necessary for all cells? Why?

7. What would happen if the blood couldn't get to the cells?

8. What body systems are involved in this activity? Give examples of parts of each system.

Righty or Lefty

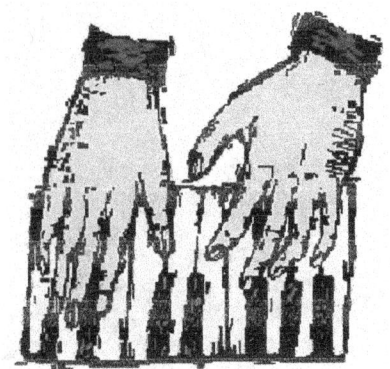

Objectives:
   Students will observe the physical characteristics used in classification.
   Students will devise a classification system.

Materials:
   Students, pencil, paper, one seating chart for each class

Procedure:
1. Your teacher has divided the class into two groups.
2. Observe the two groups and ask questions regarding the characteristics used to classify the students.

Results:
1. What characteristic was used to divide the group?

2. Why is the ability to divide individuals into groups an important skill?

3. How is this technique used in science?

4. How do you use this technique in your daily life?

5. Devise a classification system using the students in your class. Write down the characteristics you used.

Righty or Lefty
Teacher's Instructions

Objectives:

Students will observe the physical characteristics used in classification.

Students will devise a classification system.

Materials:

Students, pencil, paper, one seating chart for each class

Procedure:

1. Ask the students to write their definition of science on a piece of paper.
2. As the students are writing, note on your seating chart which are left handed. If there are no left handed students, note the positions pencils are held, type of notebook paper, arrangement of materials at their desks, or some other easily observable characteristic.
3. Divide the students into two groups.
4. Ask students to hypothesize reasons for the method of group division.

Shape Up!

Objectives:

Students will recognize shapes common to animals and to man.

Students will state the relationship between physical function and anatomical design.

Materials:

Mammals, reptiles, students, paper, pencil

Procedure:

1. Trace your hand on to a sheet of paper. Draw in some lines where you think bones would be.
2. Look closely at the front foot of a mammal.
3. Draw the outline of that foot next to the drawing of your hand. Where do you think the bones would be? Draw them in.
4. Do the same thing for the other animals.

Results:

1. Do all animals have the same number of toes or fingers?

2. Which animals have the most toes or fingers?

3. Which animals have a front foot most like your hand?

4. Why would your hand need to be different from an animal's front foot?

Skin or Scales

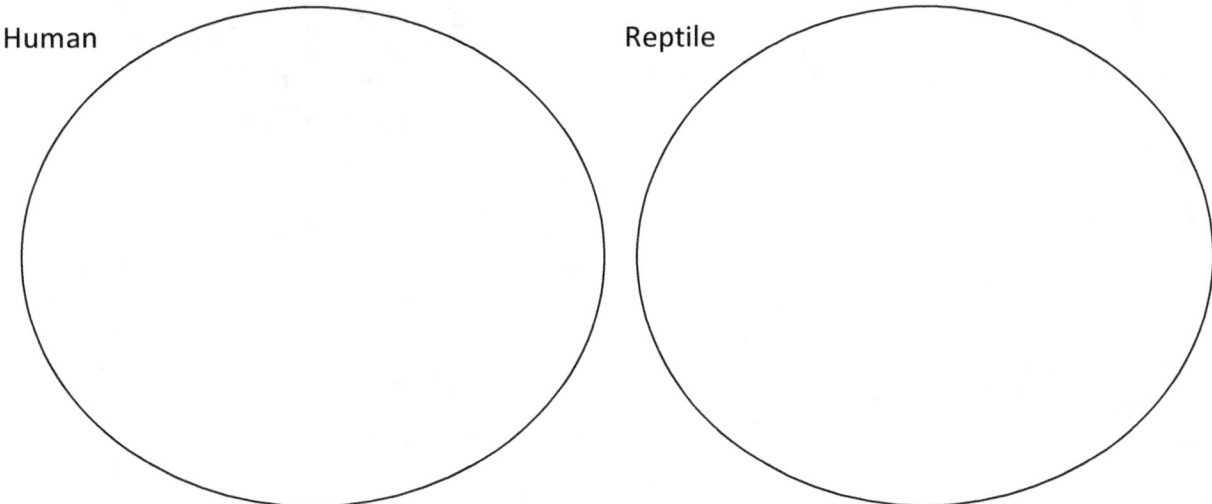

Objectives:

    Students will compare the shape of scales to the patterns in the skin of humans.

    Students will infer the function of different skin textures.

Materials:

    Hand lens, reptile, paper, pencil

Follow your teacher's directions for working with the classroom animals. Handle all animals gently and respectfully! Do not squeeze the animal. Some animals may bite, especially if they are scared or mishandled. Follow your teacher's instructions carefully.

Procedure:

1. Using the hand lens closely examine the back of your hand.
2. Draw what you see.
3. Now use the lens to look a snake's skin.
4. Draw what you see.

Results:

Human   Reptile

1. How is your skin like that of a reptile's?

2. How is the reptile's skin different?

3. Which skin seems to be softer?

4. If you lived on the floor of the forest or in a tree, would you want skin like a reptile's or like a human's? Why?

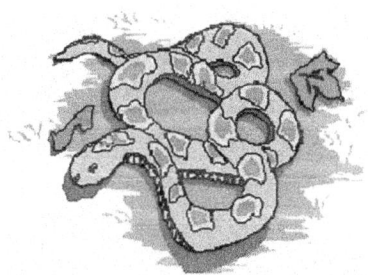

Smaller and Smaller

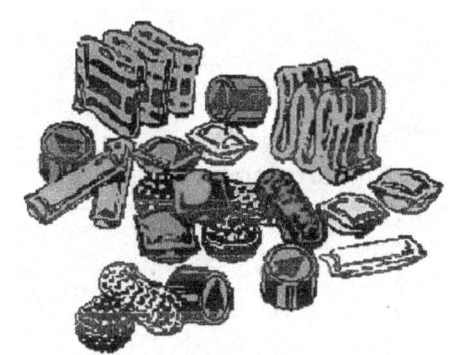

Objectives:

Students will build models of plant cells and animal cells.

Students will observe cells have smaller parts.

Students will define all living things as made of cells.

Materials:

Raw cookie dough, plastic wrap, Skittles®, large marshmallows, raisins, chocolate sprinkles, cherry licorice, Hot Tamales® candy, M&M® candy, green jelly beans, shoe box lid lined with waxed paper, oven, cookie sheets, waxed paper, small bowls, cell diagrams

Procedure:

Materials Representation

dough = cytoplasm	plastic wrap = cell membrane
Skittles® = vacuoles	marshmallows = nucleus
M&M® = ribosomes	sprinkles = chromosomes
licorice = endoplasmic reticulum	Hot Tamales® = mitochondria
jelly beans = chloroplasts	lid = cell wall
raisin (pushed into the marshmallow) = nucleolus	

1. Cover work area with waxed paper. **Wash your hands**.
2. Each pair of students gets a 6" round of dough and a bowl of cell parts. **Do not to eat any of the cell parts!!**
3. Build the plant cell first. Call out each part, refer to the diagram, and use the proper candies for each type of cell.
4. Place the cookies on a cookie sheet. Your teacher will bake the cookies following the directions on the dough package. Cool the cookies before following the rest of the directions.
5. The entire plant cell is covered in the plastic wrap. Lay the plant cell model on the lid. Follow the same procedure with the animal cell, but do not lay it on the lid.
6. Answer the questions in the Results section.
7. Dispose of the models in an appropriate manner.

Results:

1. Do you have cells? How do you know?

2. What else can you think of that has cells?

3. What are some things that are inside of cells?

4. Are animal and plant cells different? How can you tell?

5. What parts do plant cells have that animal cells do not?

6. What do you think the functions of these extra parts are?

7. Do all living things have cells? Explain.

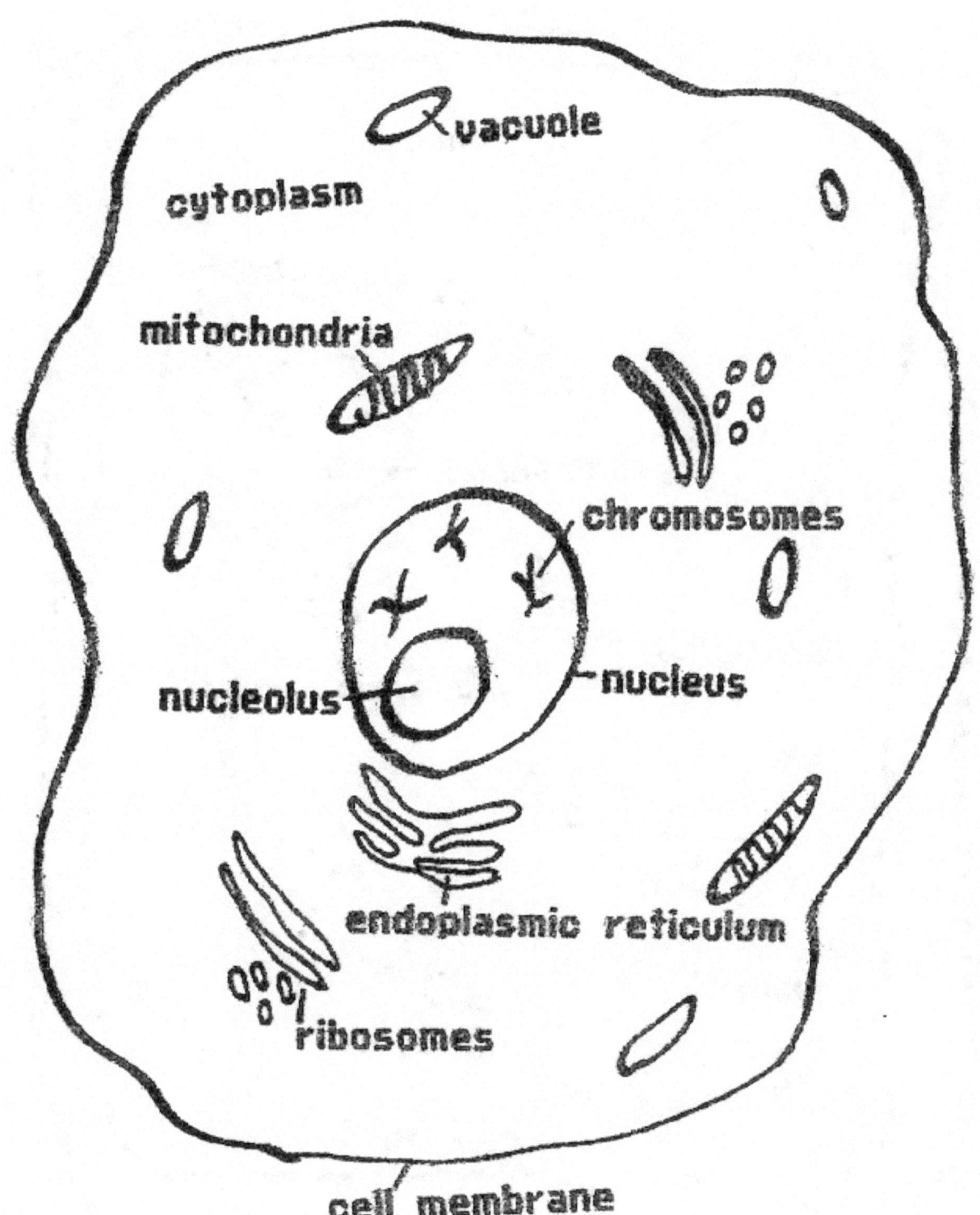

Animal Cell

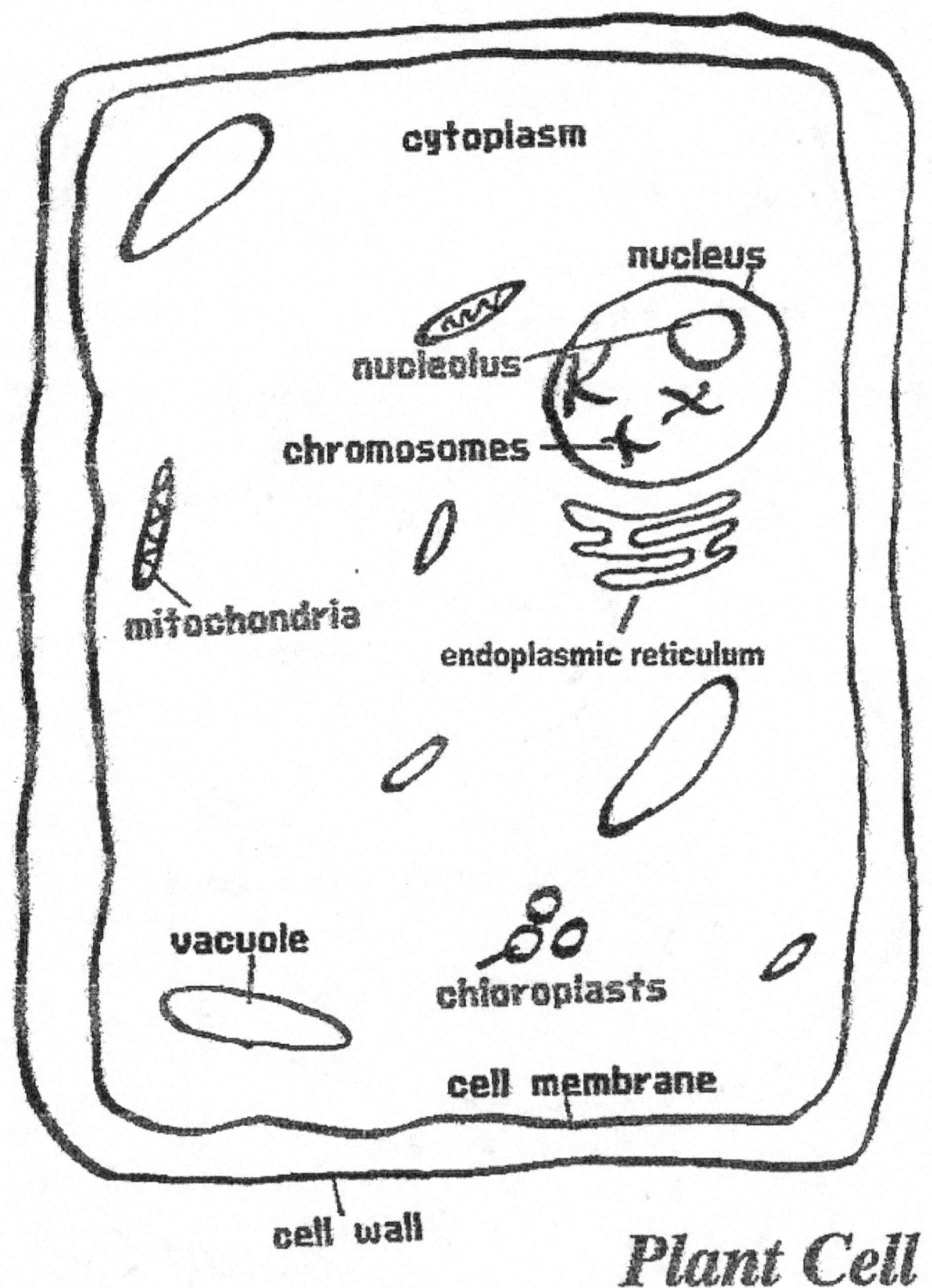

cytoplasm

nucleus

nucleolus

chromosomes

mitochondria

endoplasmic reticulum

vacuole

chloroplasts

cell membrane

cell wall

Plant Cell

Tall and Short

Objectives:
>	Students will observe the different trees in the area.
>	Students will hypothesize the reasons for the habitat in which the trees live.

Materials:
>	Paper, colors, pencils, outdoor area with trees and undergrowth

Procedures and Results:

1.	Pick a spot near the trees. What do you see when you look at this area?

2.	On a separate sheet of paper, draw some of the trees; include some of the plants under them.

3.	As you color your drawing, what do you notice about the amount of light on and under the trees?

4.	Share your picture with one other person in the class. How are the pictures alike? How are they different? What did you see that the other person did not? What did he/she see that you did not?

5.	How are the plants under the trees different from the trees?

6.	Which do you think need more sunlight, the trees or the plants under them?

7. Why do smaller plants grow under trees?

8. Which needs more food, the trees or the plants under them? Where does this food come from?

9. If you were to walk into a very dense forest, what would you see? Describe the trees and the plants under them.

What Makes It Grow?

Objectives:
    Students will manipulate variables.
    Students will predict growth rates.
    Students will determine the relationship between soil condition and plant growth.

Materials:
    Shovels, hoes, rakes, seeds, string, meter stick, bucket, large nails, signs

Procedure:
1. Go out to the tree line and collect a bucket of mulch from under the trees. Be sure you get soil and dead leaves. (Mulch may also be purchased.)
2. Measure off two 50 cm^2 areas in the garden. Put a nail in each corner of the squares and tie the string around them to form a boundary. Use the signs to label one square "A" and the other "B". (Large pots may be used if a garden area is not available.)
3. In one of the squares, turn the mulch into the soil. Turn the soil of the other square, but do not add any mulch or fertilizer.
4. Following the directions on the seed packets, plant the same types of flowers or vegetables in each square.
5. Water and weed each square regularly.

Results:
1. What happened in square "A" during the first week?

2. What happened in square "B" during the first week?

3. Which square had more weeds? Why do you think this happened?

4. In which square did the plants grow faster? Why was this true?

5. In which square did the plants grow bigger? Why did this happen?

6. What was the only difference in the two squares?

7. What did you keep the same for both squares? In an experiment, what is a control?

8. What did you change (manipulate) between the two squares?

9. What happened (responded) when you made the change?

10. What is the relationship between mulched and non-mulched soil? How do you know?

Which Way Did They Go?

Objectives:

Students will define variables as controlled, manipulated, or responding.

Students will prepare at least one experiment using materials provided by the teacher.

Materials for each pair of students:

Felt, red and green construction paper, waxed paper, cotton balls, four 3" x 5" index cards, masking tape, notebook paper, 3 sowbugs (isopods), environment for sowbugs

Procedure:

1. Tape index cards together to form a 5" square; this is your experimental environment.
2. Your job is to determine what type of environment sowbugs like best. You may use any of the materials your teacher provides for you. Remember, only change one aspect of the environment at a time.
3. Your teacher has some sowbugs in an environment. Watch these bugs carefully for at least 3 minutes before you begin your experiment. Several times during this lab, you may want to check on the bugs in your teacher's environment.
4. **Follow your teacher's directions for working with the classroom animals. Handle all animals gently and respectfully! Do not squeeze the animals**.
5. Roll the notebook paper into a funnel. Let your sowbugs slide **gently** down the funnel into the exact center of your environment.
6. Observe the sowbugs for 3 minutes. Make notes on their activities.

Results:

1. Describe the activity of the sowbugs in your teacher's environment.

2. Describe your experiment. How was it different from the environment your teacher had?

3. Describe the activity of the sowbugs in your environment.

4. What was the control? (What did you keep the same each time?)

5. What was the manipulated variable? (What did you change?)

6. What was the responding variable? (What responded to your change?)

7. What type of environment did the sowbugs seem to prefer? How do you know?

8. From your experiment hypothesize the type of environment in which sowbugs really live.

Which Way Is Up?

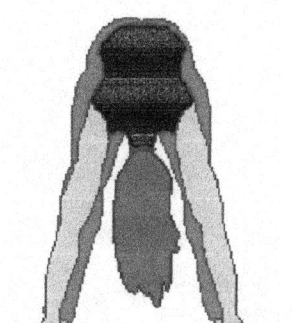

Objectives:

Students will predict the direction seeds grow.

Students will manipulate variables to determine preferred direction of growth.

Materials:

Corn seeds soaked overnight in water, clear jars with lids or petri dishes, clay, paper towels, cotton, tape, weak solution of ammonia and water

Procedure:

1. Prepare ammonia water by mixing about 15 ml of ammonia in a liter of water. The ammonia prevents growth of mold.
2. Wet the cotton and the paper towels with the ammonia water. Wring out the towels and cotton so that they are damp.
3. Lay the corn in the petri dish so that the points are at right angles (see diagram).
4. Cover them with a paper towel then pack in some cotton. You do not want the corn to slip.
5. Tape the lid on the dish. Stand the dish on its edge in some clay.
6. Observe the corn daily for a week.

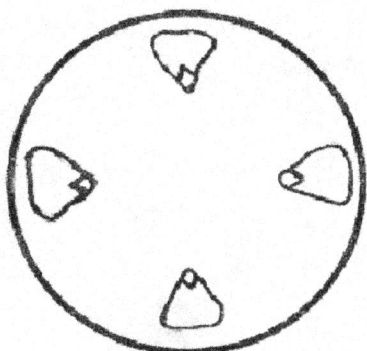

Results:

1. What happened when the corn sprouted?

2. Which way did the roots go in the beginning?

3. Which way did the stems finally go?

4. Did it matter which way the corn was pointed? How do you know?

5. What do you think controls the direction a plant grows?

6. Plant your seeds in the garden and watch them carefully.

7. Does it make a difference which way you planted a set of un-sprouted seeds? How do you know?

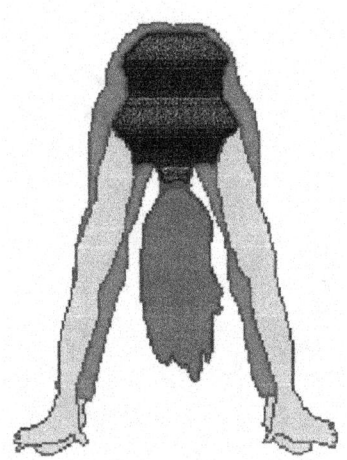

Who Lives Here?

Objectives:
   Students will observe a habitat, and decide which of the classroom animals could live in it.
   Students will infer which wild animals could be found in this habitat.

Materials:
   Paper, pencil, colors, outdoor area

Procedure and Results:
1. Sit in an outdoor area and look around you.
2. Make a quick sketch on another sheet of paper of what you see.
3. Think of the animals that live in your classroom. Which of them could live in this habitat?

4. Draw them into your sketch.
5. Think about what you see. In this habitat, is there enough food, water, and shelter for your classroom animals? Explain your answer.

6. Are there any predators in this habitat?

7. Think about wild animals. Which animals could live in the tops of the trees?

8. Do you see any evidence that supports your prediction?

9. Which animals could live at the bottom of the trees?

10. Is there water in this habitat? Which animals could live in it?

11. Would the presence of people disturb the animals in this habitat?

12. If you were an animal, what would you be?

13. Would you live in this habitat? Why or why not?

Wild Life

Objectives:
   Students will hypothesize types of habitats for the classroom animals.
   Students will state relationships between environment and body form.

Materials:
   Pencil, paper, colors, classroom animals

Procedure:

1. Look closely at the classroom animals. What kind of skin covers their bodies?

2. What do they eat?

3. When do they sleep?

4. What do their feet look like?

5. How do they use their teeth?

6. Now look at where they live. Do you think it's like their natural habitat? What is their natural home like?

7. Pretend you are one of the animals. What is your natural home like?

8. On another sheet of paper, write a story about what it is like to live in the wild. Tell about where you sleep, what you eat, and what you do all day. Illustrate your story.

9. If you would like, share your story and your drawings with the rest of the class.

10. What animal was chosen most often?

11. What animal was chosen least often?

12. Do you think you would have liked to be another of the animals more than the one you chose? Why or why not?

Physical Science

A, B, G, or X-Rays?

Objectives:
   Students will analyze data from charts.
   Students will differentiate among the three types of radiation.

Materials:
   Pencil, data sheet

Background:
   Basically, there are three types of radiation that affect humans: alpha, beta, and gamma. A thin sheet of material, such as paper can readily stop some of these rays, such as alpha (nuclei of He atoms). Beta rays are high-speed electrons. Gamma radiation is high-energy radiation. The radiation measurement most meaningful to humans is the **rem**. It is usually counted in thousandths, called millirems and abbreviated as **mrem**. This unit of measurement takes into account the biological effects of the three types of radiation. The average American receives a dose of between 620 mrem per year from all sources. (For lots of information on radiation sources, see http://www.johnstonsarchive.net/environment/lowlevelrad.html)

Procedure and Results:
1. Using the following chart, determine how much radiation you receive a year.

	Common Sources of Radiation	Annual Exposure in mrem
Where You Live	Cosmic radiation at sea level Add 1 mrem for every 100 feet of elevation. Typical elevations: Pittsburgh 1200, Minneapolis 815, Atlanta 1050, Dallas 435, Bangor 20, Spokane 1890, and Chicago 595.	30
	House construction (average) Wood 35, Concrete 45, Brick 45, Stone 50	44
	Ground (average in Texas)	18
What You Eat, Drink and Breathe	Water, food and air (average)	26
How You Live	Jet planes: Number of 6000 mile trips multiplied by 4	
	Television/Computer/Smartphone viewing: Black and white screen equals the number of hours per day. Color screen equals the number of hours per day multiplied by 2.	

	X-ray diagnosis and treatment: Check x-ray equals the number you have per year multiplied by 100. Gastrointestinal treatments equal the number you have per year multiplied by 2000. Dental x-rays equal the number you have per year multiplied by 20.	
	Average	227
How Close You Live to a Nuclear Power Plant	At the boundary: Average number of hours per day per year multiplied by 0.2 At 1 mile: Average number of hours per day per year multiplied by 0.02 At 5 miles: Average number of hours per day per year multiplied by 0.002 Over 5 miles: 0	
	Average	0
Your Exposure	Total Exposure in a Year	

2. How does your count compare with the average American?

    Higher _____ Lower _____ About the same _____

3. How might you account for any difference?

4. How close to a nuclear power plant do you live?

5. Name 3 sources of radiation that are greater than living near a nuclear power plant.

6. How could you decrease your exposure to radiation?
7. Of the types of radiation which are the most dangerous? How do you know?

8. Following is a table of types of radiation, their sources, and emission levels. Which source is greatest? Which source is least?

Type of Radiation	Some Sources	Average Annual Dose (mrem/year)
Alpha	Natural radioactivity in soils, rocks, and minerals (uranium).	30
Beta	Natural radioactivity in soils, rocks, and minerals (K-40).	20
	Television/computer/smartphone (2 hours/day average)	4
	Luminous dial wrist watch	2
	Natural radioactivity in the air (tritium)	2
Gamma	Medical and dental x-rays	50 – 100
	Cosmic radiation at sea level	40
Total		148 - 198

Action/Reaction

Objectives:

Students will apply the definition of inertia.
Students will draw conclusions.
Students will determine relationships.

Materials:

Penny, playing card, plastic cup, sheet of newspaper, thin piece of stiff wood about the size of a meter stick

Procedure and Results:

1. Balance the playing card on the plastic cup.
2. Center the penny over the mouth of the plastic cup on top of the playing card.
3. Gently slide the card off the plastic cup. What happened to the penny?

4. What will happen to the penny if you thump the edge of the card with your finger, driving it off the plastic cup top?

5. Try thumping the card. What happened to the penny? Why did this happen?

6. Place the wood flat on the table with about 12 cm sticking off the edge. Cover the wood with the newspaper, making sure it is centered over the stick and completely on the table.
7. Push down gently on the wood. Describe what happened.

8. Repeat step 6, being sure to get the newspaper absolutely flat. What do you think will happen if you slap down hard on the wood?

9. Slap the wood. What happened to the paper? What happened to the wood? Why?

10. Inertia keeps an object in motion until it is acted on by another force. Inertia also keeps an object at rest until it is acted on by another force. What was the result of inertia in step 5? What was the result of inertia in step 9?

11. You experience inertia daily. Give an example of one of these occurrences.

109

Around and Around

Objectives:

Students will observe similarities and differences.

Students will form hypotheses regarding the behavior of liquids and solids.

Students will compare and contrast the behaviors of solids and liquids.

Materials:

1 raw egg, 1 boiled egg, tabletop

Procedure for the Teacher:

1. Tell the students you have a problem; one of these eggs is raw, and one is hard boiled, but you confused them and now can't tell them apart. Without cracking the eggs, how can you decide which is which?
2. Let students make some guesses. Someone told you that a raw egg spins differently than does a boiled egg.
3. Spin both eggs for the students. [One egg will spin easily, while the other will wobble, won't spin at all, or will stop almost instantly.] Again, ask students to guess why this happens.
4. When students make a guess (hypothesis), ask them why they think they are right. [The boiled egg spins easily because it is a solid. The raw egg won't spin as well because the liquid sloshes rather than moves steadily.]

Results to Discuss:

1. Were there any visible differences in the eggs?

2. What happened when the teacher spun the eggs on the tabletop?

3. Why do you (students) think the eggs spun differently?

4. What were the reasons the eggs behaved differently?

5. How else could you tell if an egg was raw or hard boiled?

6. Compare the differences in behavior between solids and liquids.

Blowin' in the Wind

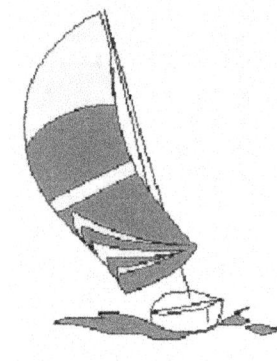

Objectives:
 Students will operationally define air pressure.
 Students will design an experiment.
 Students will predict an outcome of their experiment.
 Students will calculate air pressure.

Materials:
 3" x 5" index card, push pin, thread spool, scissors and ruler.

Procedure:
1. Insert the pushpin in the center of the index card.
2. Place the point of the pin in the hole of the spool. Do not stick the pin into the spool or your finger!
3. Hold the pin in place while blowing through the spool.
4. Continue to blow; remove your hand.

Results:
1. What happened when you removed your hand?

2. Why did the card stay on the spool?

3. Calculate the air pressure on the side of the card away from your face. Multiply the length of the card by its width. Multiply this number by 1000 g/cm^2.

4. Would you be able to blow the card off if it were smaller?

5. Design and carry out an experiment to find out how small the card must be to be blown off the spool. Record the steps for this experiment.

Bubbles and Balls

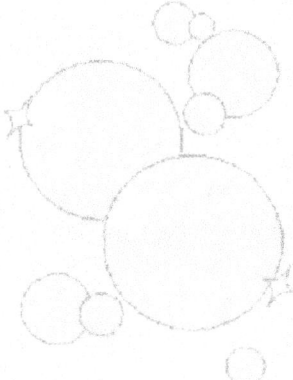

Objectives:

    Students will compare bubbles to balls.

    Students will observe properties of a gas.

    Students will analyze the relationships among air, bubbles and balls.

Materials:

    Commercially prepared bubble liquid, bubble wand, ping-pong ball.

Procedures and Results:

1. Blow bubbles and observe their direction of flight. Describe the shape, size, and color of the bubbles.

2. What happens when the bubbles touch the ground?

3. About how long does it take the bubbles to fall to the floor?

4. Hold the ping-pong ball in the palm of your hand. Can you blow the ball out of your hand?

5. Compare the ping-pong ball to the bubble.

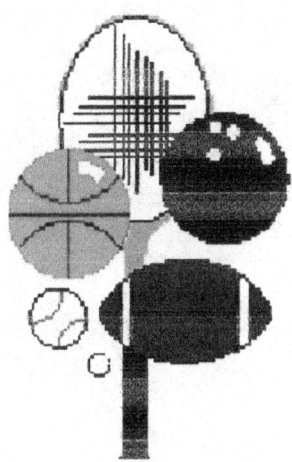

6. About how long does it take the ball to fall to the floor?

7. How does this time compare to that of the bubble?

8. If it's true that all objects are affected by gravity and fall with the same velocity, why is there a difference between the ball and the bubble?

9. The bubbles and the balls are heavier than air. Why is there a difference in the rate at which each falls?

10. What is the relationship among air, bubbles and a ping-pong ball?

114

Candle Games

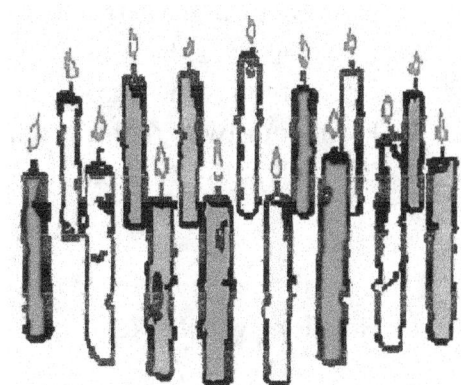

Objectives:

 Students will hypothesize reasons for the changes in the candle.

 Students will predict the rate of changes in the candle.

 Students will graph the physical properties and changes in the candle.

 Students will calculate rate of changes in the candle.

 Students will analyze types of changes in the candle.

Materials:

 Candle, metric ruler, pan or triple beam balance, graph paper, colored pencils, waxed paper, matches, stop watch or clock with a second hand, safety goggles

Procedure:

1. Find the length and mass of a new candle. Be sure to find the mass of the candle with the waxed paper under it. Record these physical properties.
2. **Put on your safety goggles**. Set the candle upright on the waxed paper, making sure that the candle cannot fall over. Light the candle.
3. Wait 2 minutes then extinguish the flame. MAKE SURE THE WAX AND WICK ARE COOL BEFORE MAKING MEASUREMENTS!
4. Without removing it from the waxed paper, measure and mass the candle. Record your results.
5. Repeat steps 2 and 3 eight more times.
6. Calculate the change in length (Δ length) and the change in mass (Δ mass); record these data on the chart.

Results:

	Length	Mass	Δ length	Δ mass
Unburned				
2 minutes				
4 minutes				
6 minutes				
8 minutes				
10 minutes				
12 minutes				
14 minutes				
16 minutes				

1. Using graph paper, plot the length in red and the mass in blue. Time should be the X-axis; physical properties the Y-axis.
2. On a separate sheet of graph paper, plot the Δ length in orange and the Δ mass in purple. Again, time is the X-axis.
3. From your two graphs, predict what the mass of the candle at the end of 24 minutes. What will its length be?

4. How long will it take for the candle to completely disappear? How do you know?

5. How is the mass related to the length of the candle? Was there a change in rate of this relationship? Explain.

6. If the candle were twice the size of the original, how long would it take to burn up?

7. Since matter can be neither created nor destroyed, what happened to the candle?

8. What parts of the candle physically changed? How do you know?

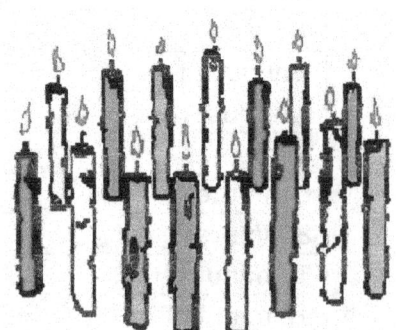

9. What parts of the candle chemically changed? How do you know?

10. You have been sentenced to study hall for four days. How many candles would you burn if you needed light for 6 hours per day?

Circuits, Circuits

Objectives:

Students will construct series circuits and parallel circuits.

Students will compare and contrast series and parallel circuits.

Students will apply their knowledge of electrical circuits to an everyday problem.

Materials:

Light bulbs, light sockets, wire, 1.5-volt dry cell, battery holder (optional), wire strippers

Procedure and Results:

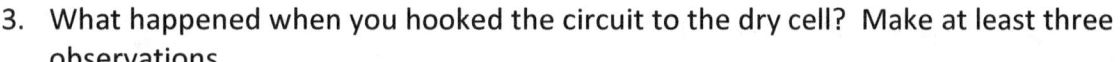

1. Cut several strips of wire about 10 cm long; strip off 1cm of insulation from each end. Wire ends may be sharp; watch your fingers!
2. Insert the light bulbs in the light sockets and wire the circuits as illustrated. This is a series circuit. From your model, why do you think this is called a series circuit?

3. What happened when you hooked the circuit to the dry cell? Make at least three observations.

4. Loosen one of the bulbs (make sure the bulb is not hot). What happens? Why do you think this is the case?

5. Rewire the circuit as illustrated. This is a parallel circuit. From your model, why do you think this is called a parallel circuit?

6. What happened when you hooked the circuit to the dry cell? Make at least three observations.

7. What differences did you notice between the first circuit and the second? Did the bulbs light differently?

8. What happens if you loosen one of the bulbs? What is happening in your circuit?

9. What is the role of the bulb in this circuit?

10. How would you wire a light switch in your house, series or parallel? Why?

11. What are some other uses for series and parallel circuits?

Davids and Goliaths

Objectives:
 Students will demonstrate the mechanical advantage of pulleys.
 Students will describe the parts of a system.
 Students will build a model of a pulley.

Materials:
 2-dowels (50 cm x 3 cm), 5 m of nylon rope

Procedure and Results:
1. Select two students and give each a dowel rod. Tie one end of the nylon rope to one of the dowels.
2. Have the students hold the ends of the dowels.
3. String the rope around the free dowel.
4. Have another student pull the loose end of the rope. While this student pulls, have the other two students holding the dowels, resist being pulled together. What happens?

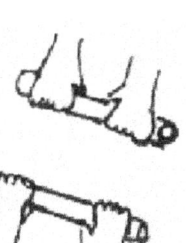

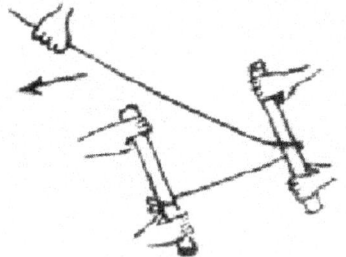

5. Loop the rope around the dowel closest to the loose end of the rope. Have the student, once again, pull the loose end of the rope. What happens?

6. Loop the rope once again. Have the student pull the loose end. What happens?

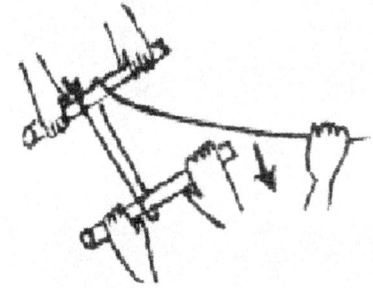

7. What is the relationship among the increase of loops of rope around the dowels, the length of the loose rope, and the efficiency of the system?

8. In a pulley system, there is a fixed pulley and a movable pulley. Which student was the fixed pulley? Did the fixed pulley change? Describe your observations.

9. What would happen to the system's efficiency if you removed the moveable pulley from the system?

10. What are the practical aspects of this phenomenon?

Fat Fingers and Skinny Fingers

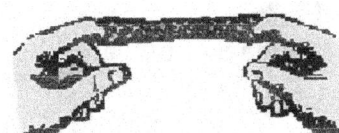

Objectives:

Students will demonstrate expansion and contraction of various objects.

Students will demonstrate heat transfer in various metals.

Students will observe changes in various materials.

Materials:

Ring and ball device (screw, eye screw, two wood handles); beaker of water; paper towels; candles; matches; aluminum pan; clothes pins; strips of aluminum, copper, nickel, brass, iron, stainless steel; safety goggles; stop watch or clock with second hand

Procedure and Results:

1. Make sure the head of the screw will not easily pass through the eye screw. Screw each into one end of separate wood handles.
2. **Put on your safety goggles.**
3. Light the candle and put one drop of wax in the center of each metal strip. Set these aside until the wax has cooled. **Keep your fingers and other flammable materials away from the flame and the hot wax.**
4. Drip a small amount of wax onto the aluminum pan and stick the candle in it to hold the candle upright. Make sure the candle will not fall over while it is burning.
5. Heat the head of the screw in the candle flame for about 1 and a half minutes. **Do not touch the screw with your fingers.** Try to pass the screw through the eye screw. What happens?

6. **The head of the screw is very hot.** Cool the head of the screw in the beaker of water and dry it with the paper towel.
7. Heat the eye screw for about 1 and a half minutes. **Do not touch the eye screw with your fingers.** Try to pass the screw head through the eye. What happens? Why does this happen?

8. What does this have to do with fingers?

9. Make sure the eye screw and screw head are separate. Cool both pieces of equipment in the beaker of water then dry them with a paper towel.

10. Predict which of the metal strips will conduct heat best. Which do you think will melt the wax first?

Metal	Prediction	Actual
Aluminum		
Copper		
Nickel		
Brass		
Iron		
Steel		

11. Hold one end of the metal in the clothespin. Put the other edge of the metal in the flame. **Keep your fingers and the clothespin away from the flame, the metal and the hot wax.**

12. Record the time it takes to melt the wax.

13. How is the first part of this activity (fitting the screw head into the eye screw) related to the second part of this activity (melting the wax)?

14. How does heat move through metal?

15. By looking at the heat conduction of the metals, which do you think would conduct electricity best? How do you know?

Floating and Sinking
Demonstration

Objectives:
 Students will hypothesize reasons for the behavior of a liquid inside a closed container.
 Students will observe relative densities.

Materials:
 1 can of diet soda, 1 can of regular soda, aquarium, water

Procedure:
1. Ask the students if the regular soft soda will sink or float in the water.
2. Drop the regular soda into the aquarium.
3. Ask the students if the diet soda will sink or float.
4. Drop the diet soda into the aquarium.

Results to Discuss:
1. How did the cans of sodas behave in the water?

2. What caused the difference in the behavior of the different sodas?

3. Density is the degree to which the atoms of a substance are packed into a given space. What is the relationship among the sodas and the water with regard to density?

4. What would happen if you shook the regular soda?

5. Try shaking the can and putting it in the aquarium. What happened?

6. Follow the same procedure with the diet soda and record your observations.

7. Why did the canned sodas behave as they did after they were shaken?

8. How is density related to the shaking of the sodas?

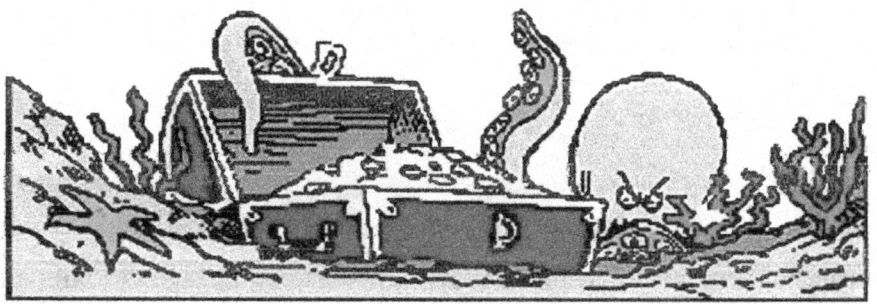

Give Me a Fulcrum – Part 1

Objectives:
> Students will demonstrate the use of levers.
> Students will observe the mechanical advantage of a lever.

Materials:
> 1 meter stick (marked and drilled at 10 cm intervals), 80 cm stick (drilled about 10 cm from one end), spring scales, string, nut and bolt

Procedure and Results:

1. Hook the spring scale under the lip of a table or desk. Pulling on the handle, see if you can lift the table or desk. If you can, record the force needed to do so.

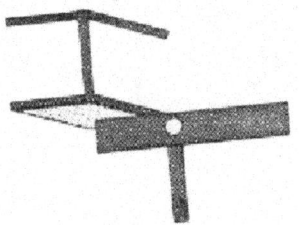

2. Attach the 1 meter stick (at the first 10 cm interval) to the 80 cm stick with the nut and bolt. The 1 meter stick is the lever; the hinge on the 80 cm stick is the fulcrum.
3. Place the 80 cm stick about 10 cm from the edge of the desk to act as the fulcrum for the 1 meter stick. Make sure the 1 meter stick is under the edge of the table or desk.
4. Push down on the 1 meter stick. What happens?

5. Try changing the length of the lever and lifting the table. How does it feel at different lengths?

Distance from the Table	Easy, Hard, Hardest
10 cm	
20 cm	
30 cm	
70 cm	

6. Try pulling down on the 1 meter stick with using the spring scale to measure the amount of effort needed. What is the reading at the following fulcrum distances?

Distance from the Table	Effort in Newtons
10 cm	
20 cm	
30 cm	
70 cm	

7. What is the relationship between the length of the lever and the mechanical advantage (effort) of using a lever?

8. List some practical uses of levers.

Give Me a Fulcrum – Part 2

Objectives:

Students will demonstrate the use of levers.

Students will observe the mechanical advantage of a lever.

Materials:

1 meter stick (marked and drilled at 10 cm intervals), 80 cm stick (drilled about 10 cm from one end), spring scales, string, nut and bolt, plastic liter bottle, wire hook and water

Procedure:

1. Bolt the center of the 1 meter stick to the 80 cm stick.
2. Fill the bottle with water and cap tightly.
3. Tie the string around the top of the bottle and tie an additional loop (See illustration). Attach the hook to the loop on the bottle.

4. Have your partner hold the 80 cm stick erect. Hook the bottle to the one meter stick at the 10 cm mark. Hook the spring scale to the 100 cm mark. Try lifting the bottle. Is it easy?

5. Hook the scale at the 90 cm mark. What is the reading on the scale?

6. Try other arrangements. Record your findings

Bottle Position	Scale Position	Reading on Scale (in Newtons)
10 cm	100 cm	
10 cm	90 cm	
20 cm	100 cm	
20 cm	90 cm	
50 cm	100 cm	
50 cm	90 cm	

7. Which direction is the bottle pulling? Which way are you exerting effort?

8. What is the relationship between the position of the scale and the position of the bottle?

9. What practical examples can you name for this arrangement?

Give Me a Fulcrum – Part 3

Objectives:

Students will demonstrate the use of levers.

Students will observe the mechanical advantage of a lever.

Materials:

1 meter stick (marked and drilled at 10 cm intervals), 80 cm stick (drilled about 10 cm from one end), spring scales, string, nut and bolt, plastic liter bottle, wire hook and water

Procedure and Results:

1. Bolt the 80 cm stick to the 1 meter stick at the ends of each stick.
2. Hook the bottle to the 100 cm mark on the 1 meter stick.
3. Hook the spring scale to the 100 cm mark. Can you lift it?

4. What are you lifting?

5. What is the force necessary to lift the bottle?

6. Try moving the scale back and record the results.

Scale Position	Scale Reading (in Newtons)
100 cm	
90 cm	
70 cm	
50 cm	
30 cm	
20 cm	

7. What happens as you move away from the bottle?

129

8.   What advantage is there to this arrangement?

9.   What are practical examples of this arrangement?

Half-Life, Whole Life?

Objectives:
Students will model rates of radioactive decay.
Students will calculate rates of decay.

Materials:
100 pennies per group, 1 shoebox with lid per group

Background:

In half-life decay, the quantity of material that is present decreases over time according to a geometric progression. In a geometric progression, the ratio between one term and the next term in the series is always the same. In the case of half-life decay, the constant ratio between successive terms is ½. After each half-life only one half as much material remains.

Procedure and Results:
1. Place 100 pennies in the shoe box heads-up. This represents all of the emission particles in the atom of this element.
2. Put the lid on the box and shake it so that the coins are well mixed. Remove from the box all coins that are heads-up. These represent the particles escaping the nucleus of the atom (i.e. decaying). Record this number on the chart below.
3. Repeat steps 2 and 3 five more times. Record the data after each "decay".

Trial	Number Heads-Up	Number Remaining	Percentage Heads-Up	Percentage of Original 100 Coins Remaining
1				
2				
3				
4				
5				
6				

4. Assuming a perfect half-life decay, what percentage of the original of coins would remain after each half-life? Answer for all 6 half-lives.

5. How do the percentages in the last column of data compare with your answers in question 6?

6. If you had the data from the entire class, would your percentages from the last column more closely match the theoretical decay pattern in question 4? How do you know?

7. If the half-life of a radioactive material were one year, how many years would have passed after 6 half-lives?

Headsium, Tailsium?

Objectives:

Students will graph rates of radioactive decay.

Students will predict rates of decay.

Students will calculate rates of decay.

Materials:

100 pennies, gallon size sealable plastic bag

Procedure:

1. You have been given a sample of the rare radioactive element Headsium (Hm) which you will use in an important medical experiment. Headsium decays over time into its daughter element, Tailsium (Tm). When a period of time known as a half-life has passed, 50% of the Hm will have decayed into Tm, releasing energy.

2. Because your sample must contain at least 20% Hm to be effective in the medical experiment, you must find out how long your sample will be good. To do this, you will model radioactive decay using the pennies.

3. A heads-up penny represents an atom of Hm, while one tails-up represents Tm. Place 100 pennies in the bag.

4. Place the plastic bag on a flat surface, such as a tabletop. Reach inside and make sure all pennies are turned heads-up. Seal the bag, making sure there is some air sealed in to give the pennies some space.

5. Shake the bag up and down ten times, trying to keep the sides of the bag parallel to the tabletop. Replace the bag on the tabletop.

6. Carefully open the bag and remove all the pennies that are tails-up. Record the number of Tailsium atoms in the data chart marked Decay Period 1.

7. Repeat this procedure for three more shakings, removing all the Tm after each "decay". Complete the data table.

8. Plot your "decays" on the graph provided.

Decay Period	Number of Tm Atoms
1	
2	
3	
4	

Number	100				
	90				
	80				
	70				
	60				
Of	50				
	40				
Atoms	30				
	20				
Left	10				
	0				
		1	2	3	4

Number   of   Decays

Results:

1. How many decay periods did it take for 50% of the Hm to decay?

2. If one decay period represented one day, what is the half-life of Hm?

3. How many days would it take for you to reach the minimum of 20% Hm in the sample?

4. What is the relationship between nuclear energy and radioactive decay?

5. For what else is radioactive decay used?

134

Hercules Unchained

Objectives:

    Students will construct an electromagnet.

    Students will compare the field strengths of electromagnets.

    Students will construct a model.

    Students will make predictions.

    Students will graph electromagnetic force.

    Students will state relationships.

Materials:

    16 penny nail, 100 cm bell wire, about 25 very small washers, emery board or sand paper, D-cell, battery holder

Procedure:

1. Scrape the insulation off the ends of the wire with the emery board.
2. Wind the wire around the nail for 20 wraps.
3. Hook the end wires up to the battery holder so that the wires contact the D-cell. **The electromagnet can get hot.** If this happens, remove the D-cell and wait for the electromagnet to cool before proceeding. **Do not burn your fingers.**
4. Predict the number of washers you can lift with the electromagnet.
5. Lift as many washers as you can with the electromagnet. Record your results.
6. Repeat steps 2, 3, 4 and 5 for 30, 40, and 50 wraps.

Results:

1. Record your prediction for 20 wraps, then lift the washers. Repeat this procedure for the other wraps, **always predicting first**.

Number of Wraps	Predicted Number of Washers	Actual Number of Washers
20		
30		
40		
50		

2. Plot your actual data on the graph.

Number

Of

Washers

20	30	40	50

Number   of   Wraps

3. What is the relationship between the number of wraps and the strength of the electromagnet?

4. What would you predict for the following number of wraps?

Wraps	Washers Lifted
15	
25	
35	
45	
55	
100	

5. How could we test these predictions?

6. Add another D-cell in parallel and then in series and record your findings.

Wraps	Parallel: Washers Lifted	Series: Washers Lifted

7. How does the strength of the electromagnet with one battery compare to the strength of the electromagnets with two D-cells attached in parallel? Attached in series?

8. In general, how does the parallel electromagnet compare to the series electromagnet?

9. How is an electromagnet different from a "regular" magnet? What would be the advantage of using an electromagnet?

Hidden Colors

Objectives:
    Students will demonstrate the difference between mixtures and solutions.
    Students will observe osmosis.

Materials:
    Filter paper (cut into 4 cm X 15 cm strips), black washable ink pen, beaker, paperclip, metric ruler, scissors, water

Procedure and Results:

1. Cut the filter paper to the size specified above. You may want to cut one end into a point.
2. Bend open the paperclip to make a bridge across the top of the beaker. Weave the paperclip through the filter paper.
3. Using the black washable ink pen, put a small dot about 1 cm from the bottom or pointed end of the filter paper.
4. Hang the filter paper and paperclip bridge into beaker. The tip of the paper should be about 1 cm from the bottom of the beaker. After you have it hanging vertically, and not touching either side of the container, remove your set up.
5. Add water to the beaker carefully. You need just enough to barely touch the tip of the filter paper.
6. Allow the water to climb up the filter paper for ten to fifteen minutes. What happens?

7. What color separated first? Last?

8. Remove the strip from the water, pour out the water and replace the bridge on the beaker. Allow it to hang and dry.
9. Do you think, now that the paper is dry, you could cut out these colors separately and add a little water to make black ink again?

10. Try it. Did it work? What happened?

11. Solutions are mixtures in which all the particles are the same size. Is the ink a solution? How do you know?

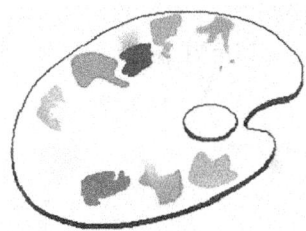

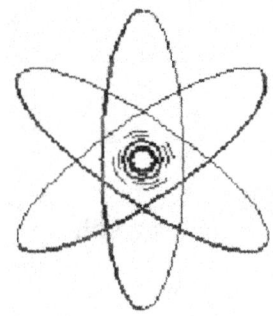

The Hysterical Atom

Objectives:

Students will operationally define fission and fusion.

Students will explain the use of a control rod in nuclear reactions.

Students will build a model of a nuclear reactor.

Materials:

25 light colored marbles (all the same size), 25 dark colored marbles (all the same size), 1 large marble, ring (such as a Hula Hoop®), inclined surface, hot plate, 2 beakers, water, water with food color added, 3 sets of dominoes, meter stick, eye dropper, safety goggles, hot mitts

Procedure A and Results:

1. Arrange the 50 marbles in the ring so that they are touching each other.
2. Set up the inclined surface such that the large marble can roll down and be dropped on the center of the marbles.
3. What happens when the large marble hits the other marbles?

4. What do the two different colors of marbles represent?

5. What does the large marble represent?

Procedure B and Results:

1. Turn on the hot plate. **Put on your safety goggles.** Half fill a beaker with plain water and set it on the hot plate to warm (do not boil). In another beaker mix water and food color together.
2. Once the plain water is warm, turn off the hot plate and **use the hot mitts** to remove the beaker from the hot plate. **Caution: the surface of the hot plate is very hot**. Use the eye dropper to place a drop of colored water on the surface of the warm water. What do you observe?

Procedure C and Results:

1. Arrange the dominoes into an ever-increasing set of rows (first row = 1, second row = 2, third row = 3, and so on until you have made a triangular shape).
2. Push the first domino. What happens? What do the dominoes represent?

3. Reset the dominoes as in step one. This time place the meter stick (control rod) between the fourth and fifth rows.
4. Push the first domino. What happens?

5. What affect did the control rod have on the reaction?

6. What might happen if you inserted the control rod in another slot between two rows?

7. What do the models in Procedures A, B, and C have to do with nuclear reactions?

8. Which of the procedures represent fusion? How do you know?

9. Which of the procedures represent fission? How do you know?

10. Why is a control rod important in a nuclear reaction?

I'll Huff and I'll Puff . . .

Objectives:

Students will operationally define air pressure.

Students will determine the relationship between the behavior of the candle flame and air pressure.

Materials:

Funnel, candle, matches, safety goggles

Procedure:

1. **Put on your safety goggles.** Light the candle. Wax from the candle may be hot; **keep the wax and the flame away from your fingers.**
2. Holding the candle at approximately 15 cm from your mouth, try to blow it out.
3. Re-light the candle. Try to extinguish it again by blowing through the small end of the funnel.

Results:

1. Could you blow out the candle without using the funnel?

2. Could you blow out the candle using the funnel?

3. What happened to the flame when you blew through the funnel? What is making the flame behave as it does? (You and your lab partner may have to repeat step 3 of the procedure several times to observe the flame.)

4. Draw the direction your breath takes as it leaves the small end of the funnel. What is happening to the air in front of the flame and the air behind the flame?

5. What is air pressure and what does it have to do with this lab?

Isaac 1

Objectives:

Students will demonstrate Newton's first law of motion--inertia.

Students will build a model of a centripetal force device.

All bodies persist in a state of rest or of uniform motion in a straight line unless compelled by some external force to change that state.

Materials:

Washer, string, 2 liter plastic bottle, cork, water, safety goggles

Procedure and Results:

1. Tie 40 cm of string to the washer.
2. Tie a loop in the other end of the string.
3. **Put on your safety goggles.** Place the loop over your index and middle fingers to secure a grip on the string.
4. Slowly begin to twirl the string around above your head, gradually increasing the speed at which you spin it. **Do not let go of the string.**
5. What happens to the washer as you increase the speed?

6. What happens to the washer as you slow it down?

7. As you spin the washer around, inertia tries to keep the washer going in a straight line, while the string pulls it back toward the center (centripetal force).
8. What is the relationship between the speed of the washer and the force you must use to spin it around?

9. Take another piece of string as long as the two liter bottle and attach the cork to it. Put the cork into the bottle and hold the string. Fill the bottle with water; screw on the cap with a small amount of string left outside the bottle. The water may leak out slightly when you invert the bottle. Hold the bottle upside down making sure the cork is floating upright.

10. Hold the bottle in both hands in front of you. Spin your entire body around in a tight circle. What happens to the cork?

11. Why do you think this happens?

12. In this model, what is happening to demonstrate centripetal force? What is happening to demonstrate inertia?

13. Give an example of a situation in which inertia is helpful. Give an example of a situation in which inertia is harmful.

Isaac 2

Objectives:
>   Students will demonstrate Newton's second law of motion--momentum.
>   Students will build a momentum device.

The rate of change of momentum is proportional to the force causing it and acts in the direction of the force.

Materials:
>   Craft stick, 3" x 5" index card, stapler, 2 coins (pennies or dimes work best), safety goggles

Procedure and Results:

1. Assemble the momentum device (see illustration) by folding the index card along its length.
    *   Tear the card along this fold.
    *   Fold the card in half along its width.
    *   Fold the two free ends to the inside center of the card.
    *   Insert the craft stick in this fold.
    *   Staple through the index card and through the craft stick.
    *   Fold the card flaps up to form a platform on either side of the craft stick.
    *   Set one coin on each platform.
2. **Put on your safety goggles.**
3. Practice using the device by putting the coins on the platforms, pulling back gently on the tip of the craft stick while holding the other end of the stick, then releasing the tip. This works a little like a slingshot.
4. Try this again, using different amounts of force to activate the momentum device. **Take care not to allow the coins to hit any other student.** What happens to the coin on the back platform (the one closest to you) when the device is activated?

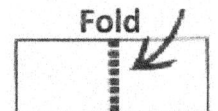

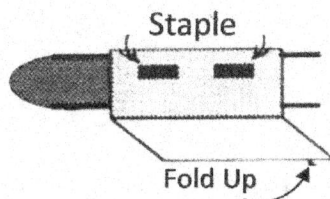

5. What happens to the coin on the front platform when the device is activated?

6. Is the distance of coin on the back platform to the floor constant? If so, what can you say about the relationship of the coin on the back platform to the floor versus the coin on the front platform to the floor?

7. Draw the forces acting on both coins. Is there any force that is common to both coins?

8. What would happen if you had a car at point C and it hit a car at point D? Circle the result of this collision.

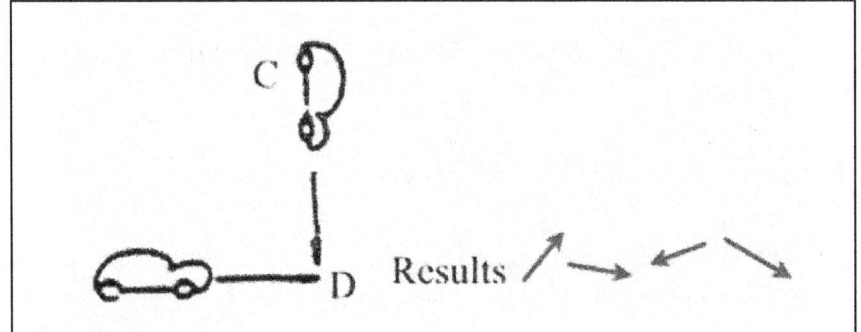

9. What do you need to know to order to determine the answer?

10. How could you illustrate your answer?

11. What does gravity have to do with this experiment?

Isaac 3

Objectives:
Students will demonstrate Newton's third law of motion--interaction.
Students will model the flight of a rocket.
Students will test variables.

If a force acts on an object, that object reacts with equal force.

Materials:
Long balloons, string, tape, straws, plastic champagne glass bases with a small hole drilled in the center of the bases, round balloon

Procedure and Results:
1. Inflate a long balloon and release it. What happens?

2. Make the long balloon hover in the air without holding on to the mouth of the balloon.
   How did you do it?

3. Attach a string to a chair on one side of the classroom. Inflate the long balloon and tape it to a straw. Thread the string through the straw. Stretch the string across the classroom.
4. Release the balloon. What happened?

5. Try again, only this time begin to vary the angle you hold the string. What happens?

6. What are some of the variables you must control to maximize the flight of the balloon?

7. What makes this experiment work?

8. Inflate a round balloon and attach it to the stem. Quickly put your finger over the hole in the bottom of the stem.
9. Remove your finger and set the base on a flat surface. What happens?

10. This type of device is the basis for a hovercraft or an air puck. Why is a hovercraft such a unique mode of transportation?

11. How do hovercraft and rockets use Newton's third law? How can you tell?

Leaky Buckets

Objectives:

    Students will explain the effects of air pressure on a fluid.

    Students will predict the various behaviors of the water.

    Students will graph data.

    Students will determine the relationship between air and water pressure.

Materials:

    Plastic two-liter bottle, masking tape, nail or ice pick, pan or sink, graduated cylinder, watch (with secondhand), ruler.

Procedure and Results:

1. With the nail or ice pick your teacher will punch three holes in the plastic bottle or equal size (diameter). Also make sure the holes are equal distant as they go up the bottle (See diagram).
2. Put masking tape over each of the holes so that the bottle will hold water.
3. Fill the bottle with water. You may wish to use food coloring to make the water more visible.
4. Put the cap on the bottle.
5. Set the bottle on the edge of the pan or sink.
6. Before you remove the masking tape, guess which of the holes will produce the longest squirt. Your guess is:

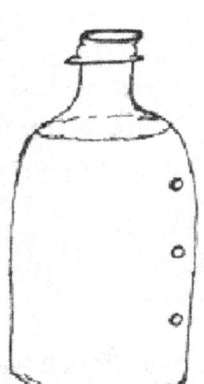

7. What happened when you removed the tape?

8. Put the tape back in place. Remove the bottle cap, then the tape.
9. Which one of the holes produced the longest squirt?

10. Was this one your guess?

149

11. Why did you have to remove the bottle cap?

12. Why did the water squirt in the manner it did?

13. Other than to empty a vessel, such as a bottle or a dam, why would we place the outlet at the bottom?

14. Refill the bottle. Re-tape the holes. Pull the tape from the bottom hole and catch the water as the water level in the bottle goes from the top hole to the bottom hole. Time this with your stopwatch. How long did it take?

15. How much water did you catch?

16. Refill the bottle. Re-tape the holes. Time the water level drop from top hole to middle hole and collect the water. How long did it take?

17. How much water did you catch?

18. Refill the bottle. Re-tape the holes. Leave the tape over the top hole. Time the water level drop from middle hole to bottom hole and catch the water. How long did it take?

19. How much water did you catch?

20. Is there a difference in the amount of water collected in a given time?

21. Graph the flow from top to middle, middle to bottom in terms of time vs. amount of water collected between holes.

Amount				
Of				
Water				
Collected				

Time   in   Seconds

22. What is the relationship between the amount of water collected and the time in seconds?

23. Is there a relationship between air pressure and water pressure? How do you know?

151

Let's Blow It Up Demonstration

Objectives:

    Students will hypothesize reasons for the behavior of a gas.

    Students test hypotheses.

    Students will apply Bernoulli's Principle to the plastic bag.

Materials:

    Garbage bag, balloon

Procedure:

1. Ask a student to inflate the balloon.
2. Have students hypothesize methods for inflating the balloon in a more efficient manner.
3. Unroll the garbage bag and ask a student to inflate it.
4. Ask students to hypothesize methods for filling the bag in a more efficient manner.

Results:

1. What is the most efficient method for filling the balloon with air? What about the garbage bag?

2. How is inflating the bag different from inflating the balloon?

3. If you pull the bag through the air, what happens to the air pressure on the outside of the bag?

4. What happens to an airplane if air pressure is reduced above its wings?

5. Bernoulli's Principle states that a quickly flowing column of air reduces the air pressure on the surface the air flows over. Does this apply to this experiment? How do you know?

Likes and Unlikes

Objectives:

Students will compare sounds of various objects.

Students will make observations.

Materials:

Tape, and 2 of each of the following: pencils, small ball of aluminum foil, 3 cm x 3 cm piece of cardboard, clothes pin, plastic spoon, washer, craft stick, clay ball, paperclip, paper grocery sacks cut a shown

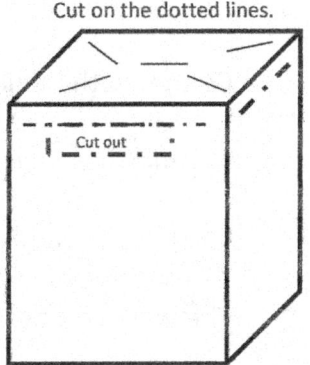

Cut on the dotted lines.

Cut out

Procedure and Results:

1. Place sacks back to back and tape them together. This is your "drop chamber".
2. Divide the objects into two groups; each partner should have the same objects.
3. Challenge your partner to a "Sound Game".
4. You drop an object on your side of the sack and then ask your partner to drop the same object on her/his side of the sack.
5. Lift the sacks to check.
6. Drop two objects at the same time. Can your partner distinguish the sounds made by the different objects?

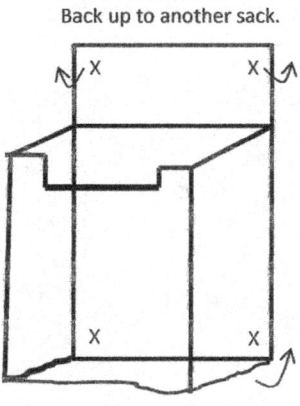

Raise bottom.
Back up to another sack.

X X

X X

Tape together at Xs.

7. Which objects sound most alike?

8. Which objects sound most different?

9. Which objects are the easiest to identify?

10. What words describe the sounds made by the dropped objects?

11. Complete the following statements:

    The plastic spoon makes a _____ sound.

    The pencil sounds like _____.

    The washer makes a _____ sound.

12. Why is it important to be able to distinguish among sounds?

13. Can we make observations with all of our senses? How do you know?

The Lion's Roar

Objective:
    Students will hypothesize reasons for the sound they hear.
    Students will observe sound waves.

Materials:
    Wooden paint stirring stick, with a hole drilled in the
handle, attached to 2.5 m of string, safety goggles

Procedure:
1. Attach the string to the stirring stick using about a 10 cm loop.
2. **Put on your safety goggles and stay away from other members of the class.**
3. Whirl the stick around your head so that it makes a low frequency hum.
4. Vary the length of the string.
5. Vary the speed with which you spin the stick.

Results:
1. What causes the sound you hear?

2. Does the sound change when the length of the string is changed? Explain.

3. Does the sound change when the spin speed is varied?

4. What musical instrument can you relate this to?

5. What would happen if the stick were tied tightly to the string rather than loosely?

A Lot of Hot Air

Objectives:

    Students will operationally define changes in air pressure.

    Students will design an experiment to test air pressure.

Materials:

    Funnel, ping-pong ball

Procedure:

1. Hold the ping-pong ball in the palm of your hand. See if you can blow it off of your hand.
2. Put the ball in the funnel. Holding the funnel above your head, blow through the small end to push out the ball.

Results:

1. What happened when you blew through the funnel?

2. How did the ball behave when you blew through the funnel?

3. Why did this happen?

4. What would happen if you held the funnel horizontally and blew very hard through it? Try this.

5. Design a method for blowing the ball out of the funnel. Report your design and your results.

Mano-Manometer!

Objectives:
    Students will construct a manometer.
    Students will form hypotheses regarding the density of several unknown liquids.

Materials:
    2 15cm pieces of glass tubing or 2 transparent soda straws, 15cm of rubber or transparent tubing, small glass funnel, 30 cm piece of rubber or transparent tubing, a piece of plastic film (approximately 20 cm x 20 cm), small ruler, large beaker (500 ml + or -), 2 pieces of wood (approximately 1" x 6" x 8", measurements given in English because these are needed at local lumber yard), water, several "unknown" liquids such as alcohol, cooking oil, glycerin

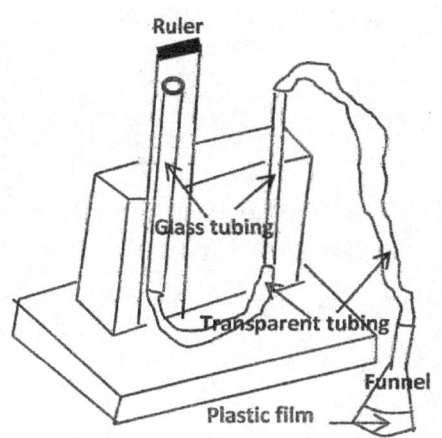

Procedures and Results:
1. Construct the manometer as illustrated.
2. Place colored water in the tube so that some can be seen on the tube side with the ruler.
3. Attach the plastic film (or membrane) to the funnel and secure with tape or rubber bands. Note the reading to which the water rises on the ruler.
4. Mark the outside of the beaker in 5 ml increments.
5. Fill the beaker with 400 ml of water.
6. Set the mouth of the funnel on the top of the water. Note the level to which the colored water rises, again.
7. Follow the procedure in steps 5 and 6 with the unknown liquids.

	Water	Unknown #1	Unknown #2	Unknown #3
First Reading				
Second Reading				

8. What does the reading on the ruler measure?

9. On which of the unknown liquids would the water float? How do you know?

10. Which of the unknown liquids would float on top of the water? How do you know?

11. If you filled the beaker again with water and also filled a large pan to the same depth with water, what would you expect the pressure readings taken from the bottom of both containers to be? Why?

12. Dense objects sink in water; less dense objects float in water. How does this information apply to this lab?

The Moor's Code?

Objectives:

Students will create a coding system using aural discrimination.

Students will compare sounds of various objects.

Students will make observations.

Materials:

Tape and 2 of each of the following: pencils, small ball of aluminum foil, 3 cm x 3 cm piece of cardboard, clothes pin, plastic spoon, washer, popsicle stick, clay ball, paper clip, paper grocery sacks cut as shown

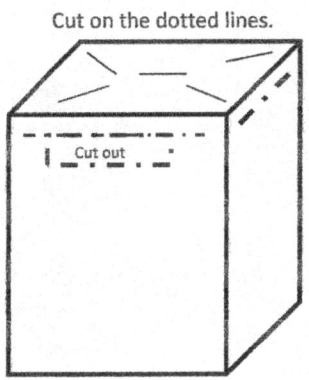

Cut on the dotted lines.

Cut out

Procedure:

1. Place sacks back to back and tape. This is your "drop chamber".
2. Divide the objects into two groups; each partner should have the same objects.
3. Challenge your partner to a "Sound Game".
   a. You drop an object on your side of the sack and then ask your partner to drop the same object on her/his side of the sack.
   b. Lift the sacks to check.
4. Select letters for use in coded messages. Suggested letters are S, T, R, E, A, M.
5. Select objects to represent these letters. For example, spoon = S. Make sure your partner knows your code.
6. Plan your message, and then send the message to your partner by dropping the objects in the sack. Take turns sending messages.
7. Exchange codes with another team and practice sending them messages. You might give them only one or two letter of your code then challenge them to break your code.
8. You can add letters to expand your code; you could even use one object as a "blank".

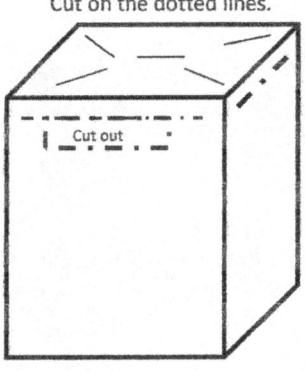

Raise bottom.
Back up to another sack.

Tape together at Xs.

Results:

1. What are the properties of the various objects that make the sounds different?

159

2. What is a practical use of sound discrimination?

3. What math messages would you send using numbers and "+", "-", "x", or ":"?

4. If you had only a spoon, how could you develop a code?

5. What guidelines would you need to make for using a code?

6. Using your objects, develop a code for this message: TEACHERS ARE GREAT!! Practice sending it to each other. What was your code?

1-500-Party Time!

Objectives:

    Students will construct a sound transmission device.

    Students will compare movement of sound waves in various media.

    Students will make a model of a telephone.

Materials:

    Plastic drinking cups with holes in the bottoms, 3 meters of string per pair of students, small washers (optional)

Procedure and Results:

1. Obtain your materials from your teacher.
2. Thread the string through the hole from the bottom and tie a small knot inside the cup. The knot should be large enough not to go through the hole when the string is stretched.
3. Holding one of the cups to your ear, have your partner talk into the other cup. What happened?

4. Try experimenting with the "telephone" you have constructed. Can you hear your partner better if he/she speaks quietly or loudly? Using a high, low, or normal voice?

5. Does the telephone work best with the string tight or loose? Will it work around a corner?

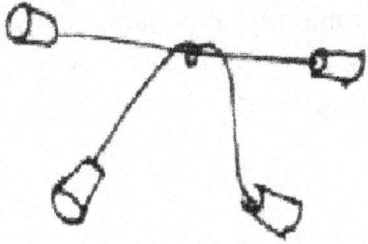

6. Try hooking your telephone to another set of partners. Have one of the new "parties" talk. Can you hear him/her?

7. How about the other people on the party line: Can they hear as well?

8. Try hooking several more strings together. Can everyone hear?

9. What is the medium you are using for the transmission of sound?

10. What are some other media you could use?

11. Why did the Indians and early trappers put their ears to the ground?

12. What other things might you hear through in the same way?

13. What are some musical instruments that transmit sound in the same manner your telephone does?

162

1-500-Party Time!!
Teacher's Instructions

Objective:

Students will construct a sound transmission device.

Students will compare movement of sound waves in various media.

Students will make a model of a telephone.

Materials:

Plastic drinking cups, 3 meters of string per pair of students, small (#6) nail, scissors, candle, safety goggles, matches

Teacher's Procedure:

1. **Put on your safety goggles.** Heat the nail over the candle flame, then push a small hole through the bottom, center of each cup. **Keep your fingers and other flammable materials away from the candle flame. The nail will get hot; exercise caution.**
2. Cut 3 meters of string for every pair of students.
3. If the holes in the cups are too large to keep the knotted string from pulling through, tie a small washer to the end of the string. This will also improve the sound quality.
4. As a variation, give two students 10 meters of string with a cup at each end. Give other pairs of students 50 cm of string and two cups. Have the students with the shorter string tie cups to both ends of the string. They then have both an ear-piece and a mouthpiece. Wrap the shorter string around the longer string.

Pepper Dolphins
Demonstration

Objectives:

Students will observe the activity of the pepper.

Students will hypothesize reasons for the behavior of the pepper.

Students will predict how the pepper will behave in the presence of another substance.

Students will determine the relationship between soap and surface tension.

Materials:

Petri dish, water, pepper, liquid detergent, sugar cube, toothpick, overhead projector (optional)

Procedure:

1. Half fill the petri dish with water. You may want to set it on top of the overhead projector and project the image. Sprinkle the pepper on the top. Tell the following story:

   As you can see, there are dolphins swimming in the ocean. Some are close together, while others are far apart. What would happen if a particularly tasty fish were in the center of this group?

2. Place the sugar cube in the center of the petri dish. After the activity stops, remove the sugar cube and continue:

   So the dolphins have eaten the fish and are now relaxing. But what do you think would happen if a killer whale came by?

3. Using the toothpick put a tiny drop of liquid detergent in the water near the edge of the petri dish.

Results:

1. Tell how the pepper acted throughout the demonstration.

2. Why did the pepper react as it did when the sugar was added?

3. What did the liquid detergent represent?

4. Surface tension is water's ability to stick together. How is surface tension involved in this experiment?

5. What is the relationship between the molecules of water and the liquid detergent? How do you know?

Ramps and Stairs

Objectives:

Students will operationally define effort, resistance, and mechanical advantage.

Students will calculate mechanical advantage from their observed data.

Students will determine the relationship between the mechanical advantage and the height of the inclined plane.

Materials:

String, spring scale, textbook, metric ruler, inclined plane

Procedure:
1. Tie the string around the length of your textbook.
2. Attach the string to the spring scale. Practice dragging the book up the inclined plane at a constant rate. Read the spring scale at the same time.
3. Weigh your textbook. Record its weight in Newtons.
4. Elevate the inclined plane to 5 cm at one end. Determine the effort needed to move the book up the ramp. Repeat this procedure for 10 cm and 20 cm. Record your data. Calculate the mechanical advantage (MA).

Results:

Record your data in the chart below.

Elevation (cm)	Weight (Newtons)	Effort (Newtons)	MA (weight/effort)
5 cm			
10 cm			
20 cm			

1. What is the relationship between the elevation of the inclined plane and the mechanical advantage?

2. What is the constant in this experiment?

3. What is the effort in this experiment? When did it change?

4. From your experiment, what is mechanic advantage?

5. What forces were making you expend energy to move the book? How could you tell these forces were acting on the book?

6. Would the length of the inclined plane make any difference? How do you know?

7. How could you determine the mechanical advantage of a set of stairs?

8. Would it be harder to move an object up a steep flight of stairs or up a gradually sloping set of stairs? Why?

Rising Water

Objectives:

    Students will predict the behavior of the water.

    Students will relate changes in air pressure to events in their lives.

Materials:

    Two bulbless medicine droppers, beaker, water

Procedure:

1. Fill a beaker three-quarters full.
2. Hold the small ends of the droppers close to each other at right angles.
3. Put the large end of one into the water and blow through the large end of the other.

Results:

1. What happened when you blew through the dropper?

2. If you blew harder what would happen to the water?

3. Why did this happen?

4. How does the phenomenon you demonstrated explain the flight of airplanes?

5. In what other circumstances is this principle used every day? (Hint: think about shower curtains.)

Roses Are Red, Violets Are Blue!

Objectives:

Students will explore reactions of materials to acid/base indicators.

Students will observe color changes in materials.

Materials:

2 small beakers or cups, some rose petals and violet petals, plastic spoon, water, vinegar, baking soda, and 2 plastic glasses.

Procedure and Results:

1. Add a small amount of water to each of the beakers.
2. Add some rose petals to one of the beakers and some violet petals to the other.
3. Using your spoon, gently crush the rose petals until they turn the water red. Repeat for the violet petals.
4. Once you have the petals crushed and the color removed from them, add a little more water to each and pour into the plastic glasses.
5. In the glass containing the red color from the rose add a small amount of baking soda and stir. What happens?

6. In the glass containing the violet color add a small amount of vinegar. What happened?

7. Add a small amount of vinegar to the rose (originally red) glass. What happened?

8. What do you think would happen if you added some baking soda to the violet glass?

9.   Try it and see. What happened?

10.   How long do you think this process could go on?

11.   What conclusions can you make about the "juice" in roses and in violets?

Round and Round, Up or Down?

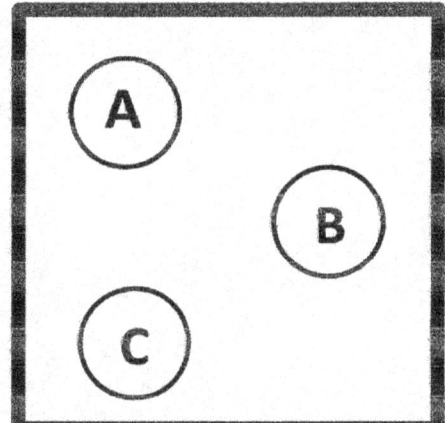

Objectives:

Students will demonstrate the use of gears.

Students will determine the relationship between band arrangement and direction of rotation.

Students will build a model of a gear system.

Students will identify methods of using mechanical advantage.

Materials:

Demonstration board, 3 spools of the same size, one larger spool, rubber bands

Procedure and Results:

1. Place the spools of equal sizes on the demo board as illustrated.
2. Place a rubber band around spools A and B. When you turn spool A clockwise, what happens to spool B?

3. Place another rubber band around spool B and C. When you turn spool A clockwise, what happens to C?

4. Remove the rubber bands. Predict how might you place the rubber band on spools A and B to make spool B turn a different direction from spool A.

5. Test your arrangement. What happened?

171

6. What happens to spool C?

7. Put a larger spool at position A. Repeat steps 2 and 3. What happens to the other spools?

8. If these spools were gears, how would using them give you a mechanical advantage?

9. Give some examples of where gears are used.

Round and Round, Up or Down?
Teacher's Instructions

Objectives:

Students will demonstrate the use of gears.

Students will determine the relationship between band arrangement and direction of rotation.

Students will build a model of a gear system.

Students will identify methods of using mechanical advantage.

Materials:

4" x 4" board, 3 1/4" dowels, spools, rubber bands, drill, glue, safety goggles

Procedures:

1. Put on your safety goggles. Drill holes at the three corners of the board just large enough for the dowels to fit.
2. Put several drops of glue into each hole.
3. Position the dowels in the holes and allow the glue to dry.

Soda Pop!

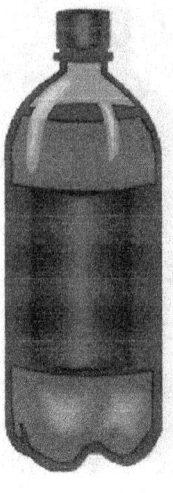

Objectives:

Students will demonstrate a chemical reaction.

Students will observe phases of materials.

Materials:

Small plastic flask (50-100 ml, or plastic .5 liter bottle), balloons, funnel, 2 plastic spoons, vinegar and baking soda, safety goggles

Procedure and Results:

1. **Put on your safety goggles.**
2. Put two spoonfuls of baking soda in the plastic bottle.
3. Insert the funnel stem into the balloon.
4. Using the second spoon put two to three spoonfuls of vinegar into the balloon through the funnel.
5. Carefully remove the funnel from the balloon.
6. Twist or pinch the balloon tightly just above the level of the vinegar.
7. Have your partner carefully spread open the end of the balloon and slip it over the top of the bottle. **CAUTION: Before you drain the vinegar into the flask, make sure the top of the flask and the balloon are pointing away from both your partner's and your faces**.
8. Un-twist the balloon allowing the vinegar to drain into the bottle containing the baking soda.
9. What happened?

10. What do you think is formed by this combining of vinegar and baking soda?

11. Watch for a minute, what happens?

12. Describe the chemical reaction.

13. Describe the different phases of matter.

Sticky Stuff

Objectives:

Students will identify cohesion and adhesion.

Students will observe a property of water.

Materials:

Small cup, water, paper large enough to cover the mouth of the cup

Procedure:

1. Dampen the lip of the cup. Usually ten drops of water is enough.
2. Place a piece of paper over the mouth of the cup.
3. Using one hand to hold the paper on the cup, invert the cup and the paper.
4. With the cup held upside down, gently twist it.
5. Remove your hand.

Results:

1. What happened when you removed your hand?

2. Why did this happen?

3. What would happen if you rub the lip of the cup with honey?

4. What property does the water and the honey share?

5. If water did not have this "sticky" property, how would it change our lives?

6. When a substance sticks to itself, it has cohesion. Does water have the property of cohesion? How do you know?

7. When a substance sticks to something else, it has adhesion. Does water have the property of adhesion? How do you know?

Test Pilots

Objectives:
    Students will compare and contrast models.
    Students will predict outcomes of certain activities.
    Students will observe the effects of air pressure on models.
    Students will design an experiment.
    Students will collect and graph data.

Materials:
    Sheets of typing paper, scissors, tape, soda straws, pencil, stopwatch, meter stick, paper clips

Procedure and Results:

1. Make a paper airplane of your own design and test it. Describe your design. You may make a drawing of your airplane.

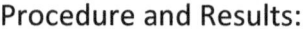

2. Keep a record of any alterations you made in the design as you tested it.

3. Once you are satisfied with your design, complete the following tests:
   A. *Duration:* Launch your plane. Using a stopwatch, determine how many seconds it flies. Repeat for ten trials.

Trials	Seconds	Trials	Seconds
1		6	
2		7	
3		8	
4		9	
5		10	

Plot your test flight duration on the graph.

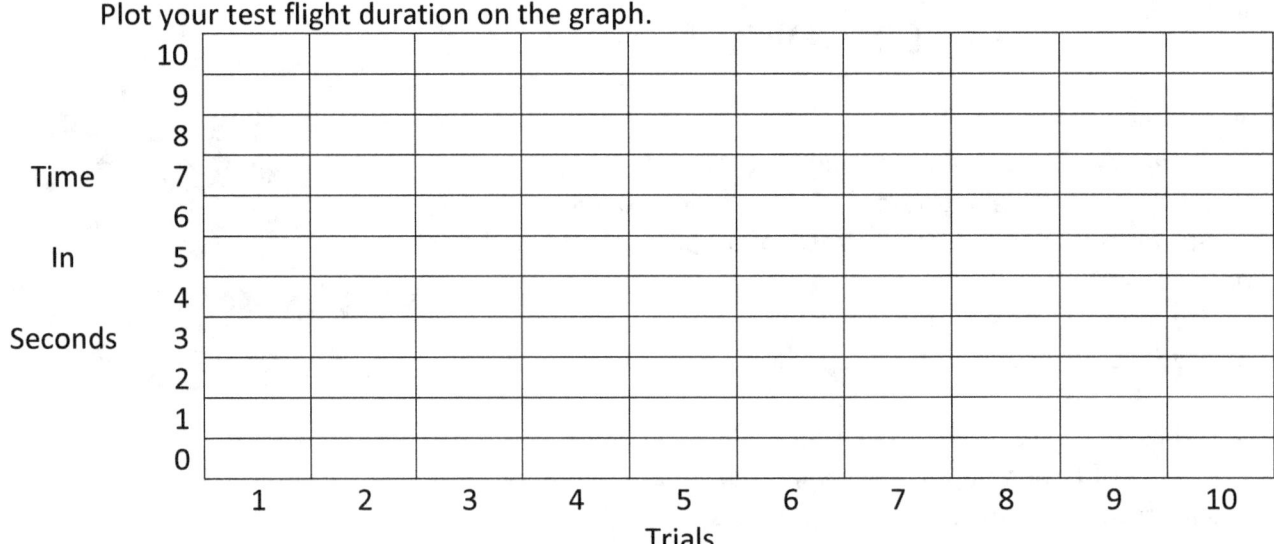

What was your average duration?

B. *Altitude:* Does you plane fly high, or does hang above the ground? Practice launching your plane, this time for altitude. Evaluate its performance as Excellent, Good or Poor. Complete ten trials.

Trials	Altitude (E, G, P)	Trials	Altitude (E, G, P)
1		6	
2		7	
3		8	
4		9	
5		10	

Plot your results on the graph.

What was your average altitude performance?

4. This time try making the airplane using two paper loops (2 cm by 12 cm) taped to opposite ends of a soda straw.

   A. *Duration:* Launch your straw. Using a stopwatch, determine how many seconds it flies. Repeat for ten trials.

Trials	Seconds	Trials	Seconds
1		6	
2		7	
3		8	
4		9	
5		10	

Plot your test flight duration on the graph.

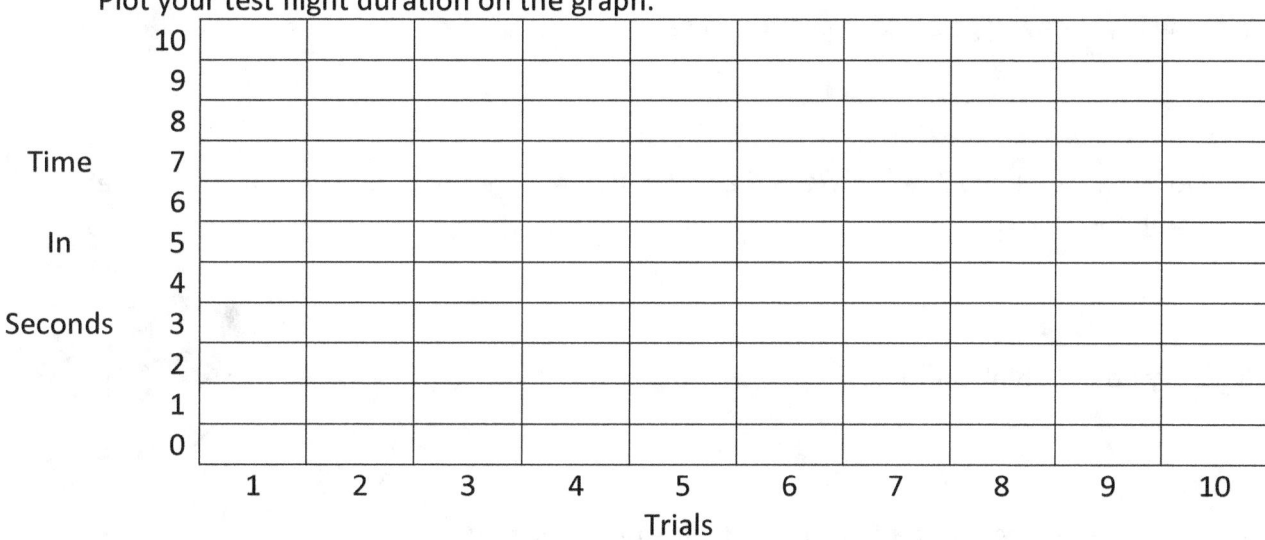

What was your average duration?

   B. *Altitude:* Does you straw fly high, or does hang above the ground? Practice launching your straw, this time for altitude. Evaluate its performance as Excellent, Good or Poor. Complete ten trials.

Trials	Altitude (E, G, P)	Trials	Altitude (E, G, P)
1		6	
2		7	
3		8	
4		9	
5		10	

Plot your results on the graph.

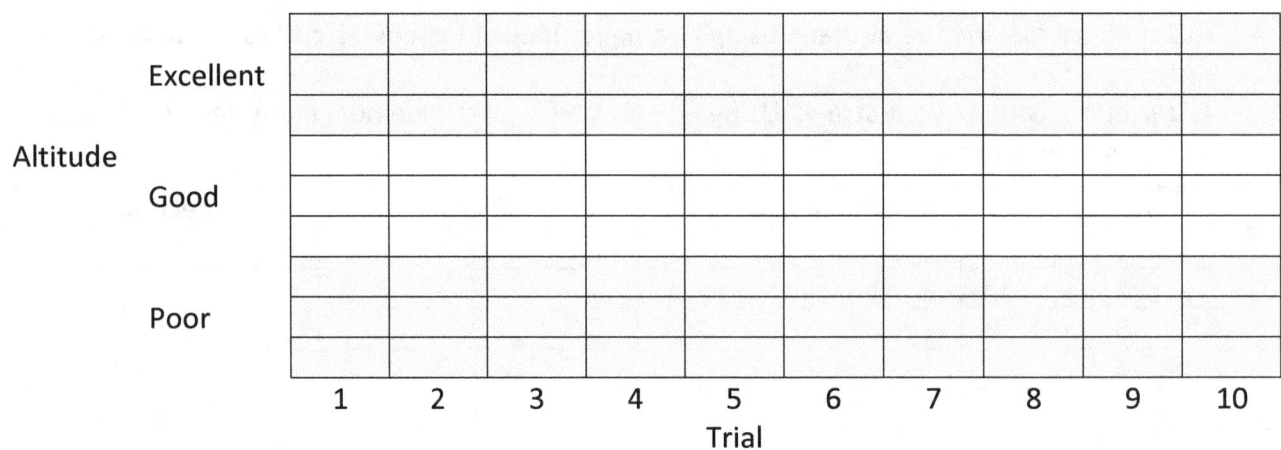

What was your average altitude performance?

5. Now that you have flown two types of aircraft, which seemed to perform better in terms of flight duration?

6. Why do you think this is so?

7. Which performed better in terms of high altitude flying?

8. Why so you think this is so?

9. What are the variables you could control to make your airplane fly better?

10. What variables could you control to make your straw fly better?

11. Fly each of your aircraft again, this time measuring distance flown.

Trials	Plane distance (cm)	Straw distance (cm)
1		
2		
3		
4		
5		
6		
7		
8		
9		
10		

12. Which craft flew farther?

13. From your previous data, which craft averaged the longer flight?

14. What do you think your data mean?

15. Design a rocket. Use the straw as your templates and launching pad. Describe your design. You may make a drawing of your airplane.

16. Keep a record of any alterations you made in the design as you tested it.

17. Record your test flights in seconds.

Trials	Seconds	Trials	Seconds
1		6	
2		7	
3		8	
4		9	
5		10	

Plot your test flight duration on the graph.

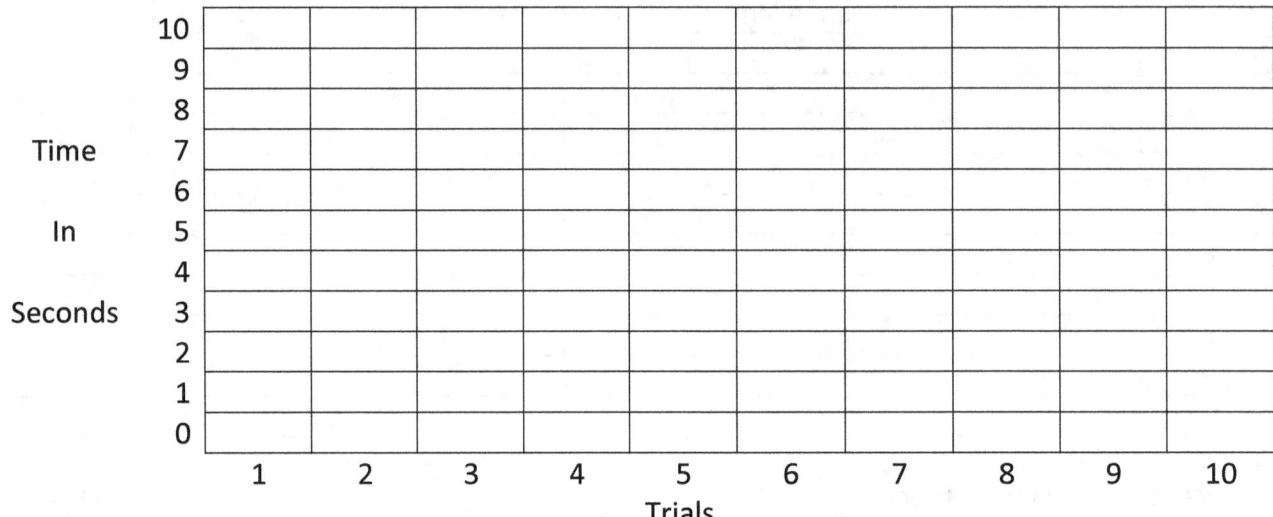

What was your average flight in seconds?

18. What is the difference in the design of your airplane and of your rocket?

19. Did your rocket fly as far as those of your classmates did? Why?

20. What are the variables to control in flying your rocket?

21. Record the following class rocket flight data:

    Longest duration

    Shortest duration

    Average duration

22. What are the variables associated with the longest flight?

23. What are the variables associated with the shortest flight?

24. Why do you have to consider variables when designing models?

25. Does air pressure have anything to do with the flight of your models? How do you know?

Test Pilots Teacher's Instructions

Objectives:
>   Students will compare and contrast models.
>   Students will predict outcomes of certain activities.
>   Students will observe the effects of air pressure on models.
>   Students will design an experiment.
>   Students will collect and graph data.

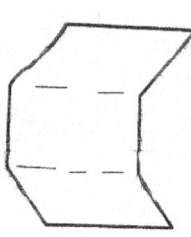

Materials:
>   Sheets of typing paper, scissors, tape, soda straws, pencil, stopwatch, meter stick, paper clips

>   Some students may need help in designing the airplane and the rocket. The Stingree can be used for the airplane tests. Instructions for a simple rocket are also included.

Procedure:
1. Fold a sheet of paper in thirds like a business letter. Crease the folds and tear off one section of the paper.
2. Roll one section around the straw at an angle to make a tube. The tube should slide easily on the straw. Use small pieces of tape to hold the tube together.
3. Cut off the tip at each end of the tube. If one end of the tube seems tighter, use this as the nose cone.
4. Cut the smaller end of the tube into 5 or 6 points as illustrated.
5. Form points into a nose cone and tape together.
6. With the remaining folded paper, trace the fin pattern.
7. Cut out the fins and attach to the tube with the tape. Fold the fins out from the tube as illustrated.
8. Place the tube over the straw and blow to launch the rocket.

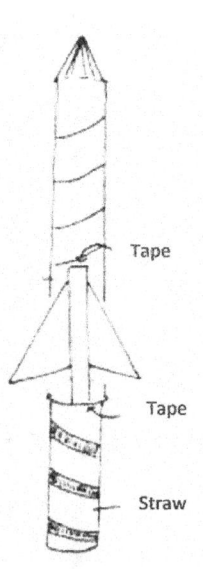

Tape

Tape

Straw

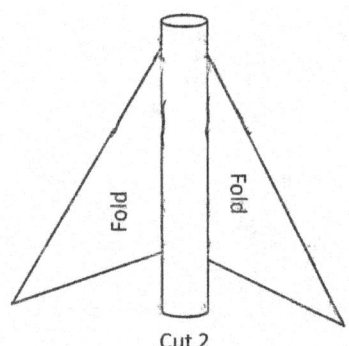

Fold   Fold

Cut 2

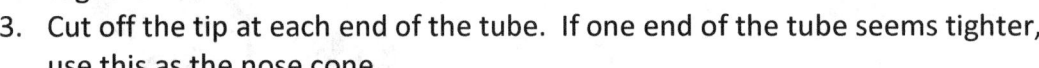

Fins - - Bottom View

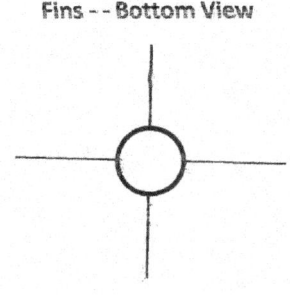

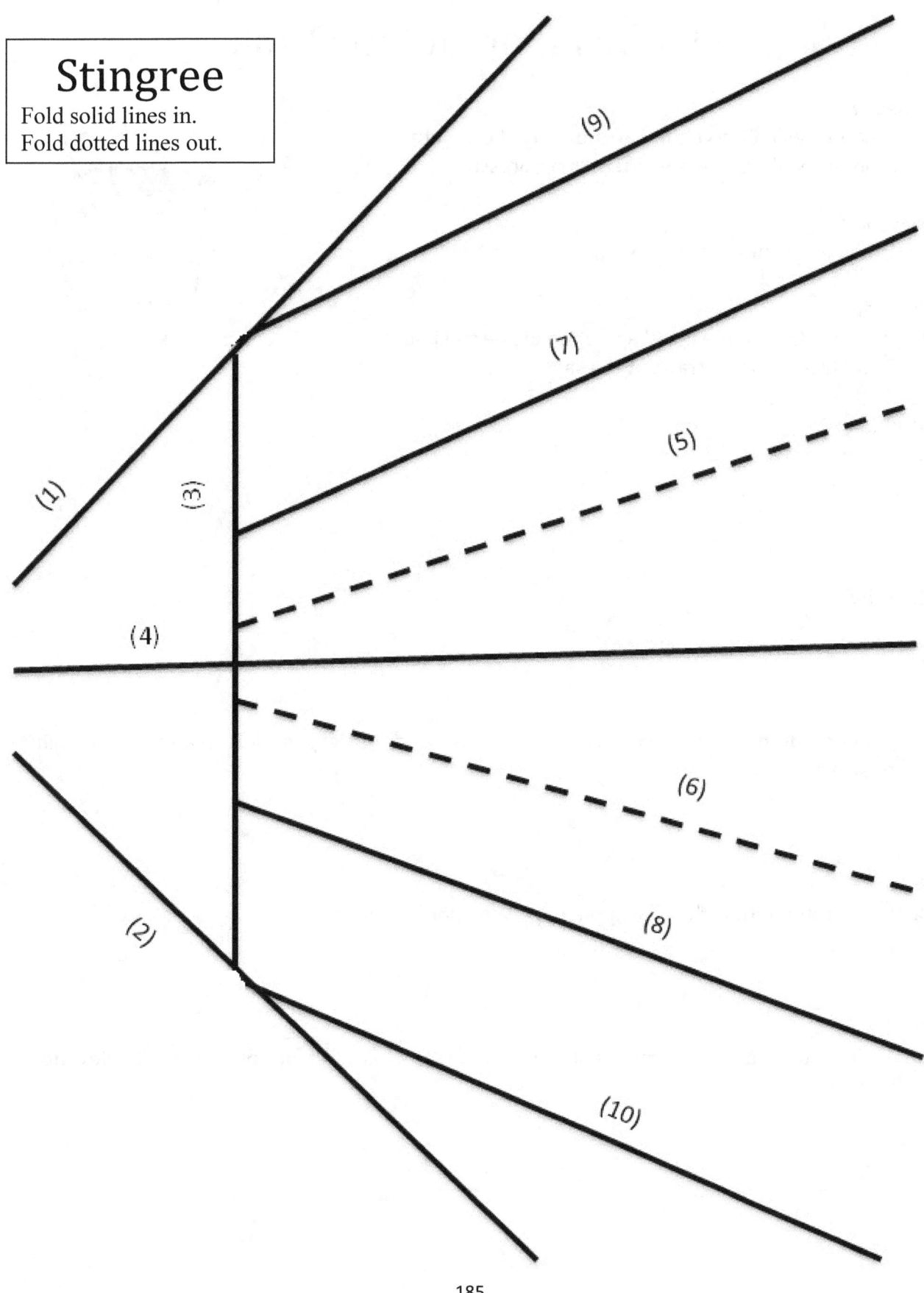

Stingree
Fold solid lines in.
Fold dotted lines out.

To Drink or Not to Drink

Objectives:

    Students will observe the effect of partial pressure.

    Students will observe the effect of air pressure.

Materials:

    Liquid in a container, 2 straws per student

Procedure:

1. Put one straw in the liquid and one outside the liquid.
2. Suck through both straws at the same time.

Results:

1. Did you get a drink?

2. Why?

3. Some children are born with a cleft pallet. How does this gap in their lip affect their ability to suckle?

4. Rabbits are born with a gap in their lip. Why don't they starve?

5. Could you get a drink through a straw if there was no air pressure on the liquid? How do you know?

Tree Limbs and Pulleys

Objectives:

Students will determine the mechanical advantage of various pulley systems.

Students will give examples of pulley systems.

Students will calculate mechanical advantage.

Materials:

Pulleys, 1 meter of string with a loop on each end, spring scales, small plastic bottle of water, dowel, duct tape

Procedure and Results:

1. Use the spring scale to find the weight of the bottle of water. Record your findings. Which direction did you have to pull to lift the bottle of water?

2. Attach the string to the water bottle by running the string through one of its loops and around the neck of the bottle. Put the other end of the string around the pulley, making sure it is in the groove.

3. Tape the dowel to a table or desk. Hook one end of the pulley to it. Is this your fixed or your movable pulley? How do you know?

4. Hook the spring scale to the loose end of the string. Pull on the spring scale, allowing the pulley to help you lift the bottle. **Did the weight of the bottle change? Note the weight of the bottle.** A single, fixed pulley offers no gain except in direction. What do you think the mechanical advantage (MA) is?

5. Try to rig the following examples and determine the mechanical advantage (MA = R/E). R is the resistance or weight of the bottle; E is the effort.

    Resistance_____

    Effort_____

    MA_____

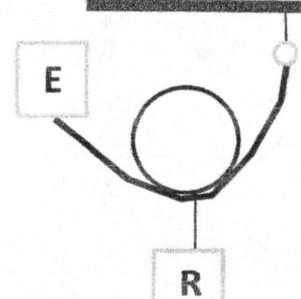

Resistance_____

Effort_____

MA_____

Resistance_____

Effort_____

MA_____

6. What is the advantage of having more than one pulley in a system?

7. What is the advantage of having a movable pulley?

8. Try to estimate the MA of the following pulley systems. Put your estimate above each.

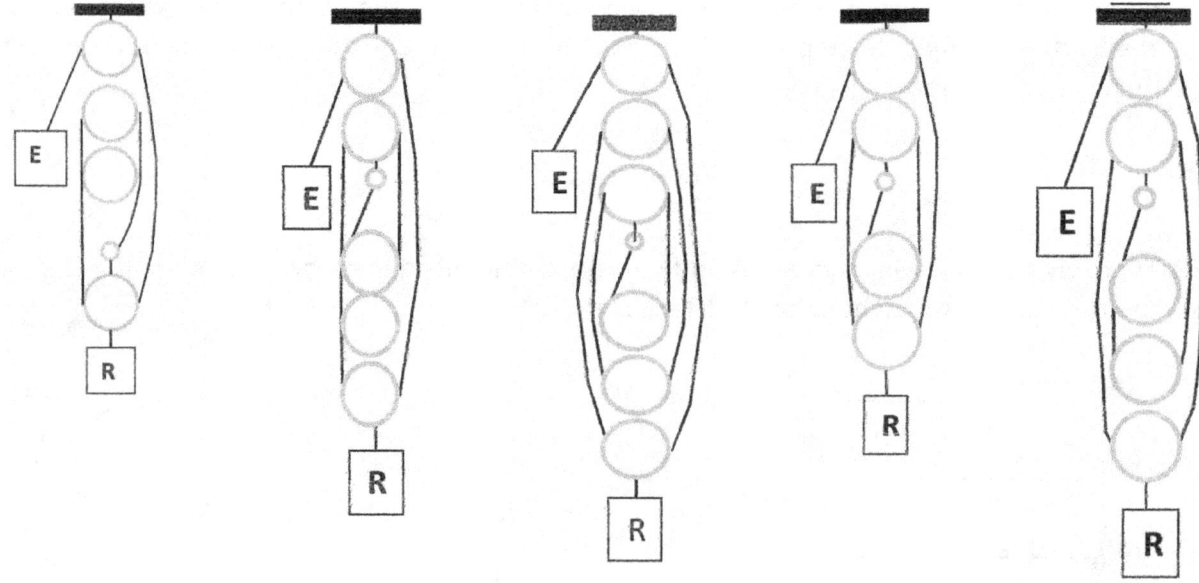

A rope over a tree limb, a set of gears, a bicycle, and a hoist are a few examples of pulleys. Name a pulley system that can be found in your home.

9. Pulleys are used to gain power, speed, and/or change direction. Give an example of a pulley that causes a change in direction

Water Race

Objectives:
    Students will hypothesize methods for removing liquids from containers.
    Students will observe the effects of air pressure on a liquid.

Materials:
    2 gallon milk jugs, aquarium, water

Procedure for the Teacher:
1. Fill both jugs with water.
2. Challenge a student to race the teacher in emptying the water into the aquarium.
3. To win, stick your finger into the mouth of the jug, allowing air to displace the water.

Results:
1. What difference did sticking your finger into the mouth of the jug make?

2. Why does water behave in this manner?

3. To make juice pour out of the can easier, what should you do to the can that doesn't involve sticking your finger in it?

A Weighty Subject

Objectives:

Students will predict the amount of force it takes to break an egg.

Students will hypothesize reasons for the varying strength of an eggshell in different positions.

Materials:

Waxed paper, textbooks of equal size and weight, 12 paper plates, clay, 12 eggs, newspaper, scale large enough to weigh a textbook

Procedure:

1. Weigh a textbook. Write the weight in the Results section. Predict the number of books that can be stacked on an egg before it cracks. Write your prediction in the Results section.
2. Cover the lab area with newspaper. Fold each paper plate in half twice; mark these diameter lines.
3. Place the paper plates on a large sheet of waxed paper, on the newspaper.
4. Roll 4 pieces of clay into 15 cm "snakes". Make rings and center them on the paper plates using the diameter marks.
5. Stand the eggs, large ends down, in the rings. The eggs should be positioned so that the books rest on all four eggs at once.
6. Cover the eggs with a large piece of waxed paper.
7. Gently begin placing the books on top of the eggs. Continue to add books until one of the eggs cracks.

Results:

1. Weight of the book:

   Prediction:

2. How many books did you put on the four eggs?

3. Multiply the number of books by the weight of one book to get the total weight.

4. Divide the total weight by 4 to find out how much weight each egg held.

5. Why can eggs hold this much weight?

6. Why is it important that an eggshell can withstand a great deal of weight?

7. What would happen if the eggs were turned on their sides rather than with the large ends down? Could they stand more, the same, or less weight?

8. Repeat steps 2 through 7 of the procedure with the eggs turned on their sides. Compare your results to your first set of data. Why did the eggs behave as they did?

9. What would happen if the eggs were turned with their small ends down? Could they stand more, the same, or less weight?

10. Repeat steps 2 through 7 of the procedure with the eggs turned with their small ends down. Compare your results to your first and second sets of data. Why did the eggs behave as they did?

11. Why must architects consider forces when designing multi-story buildings?

Where's the Fire?

Objectives:
Students will hypothesize the effects of temperature on a flame.
Students will predict the amount of heat necessary to maintain a flame.

Materials:
Candle, candle holder, matches, coiled wire (slightly finer than coat hanger wire), safety goggles, hot mitt

Procedure:
1. Coil the wire in the shape of a candle snuffer, leaving about 1 mm between each coil.
2. **Put on your safety goggles.** Mount the candle in the candle holder and light it. **Keep fingers and other flammable materials away from the flame. Use the hot mitt to hold the wire,**
3. Bring the wire down over the flame, but do not touch the wick.

Results:
1. What happened when the wire covered the flame?

2. Why did this happen?

3. What would you have to change to allow the candle to continue burning?

4. Where did the heat from the candle flame go? How do you know?

Wild Horses

Objectives:

    Students will calculate horsepower.

    Students will compare horsepower capabilities of other students.

    Students will identify energy transformations.

Materials:

    String, weight, stairs, stopwatch, meter stick, metric bathroom scale

Procedure:

1. Measure the height of the stairs by dropping a weight tied to a string from the top. Record the height in meters on the chart below.
2. Weigh your volunteer runners. Be sure to vary the size of the volunteers. Record their weights on the chart.
3. Have each volunteer run up the stairs as rapidly as possible while you time each one. Record their times.
4. Calculate the horsepower generated by each volunteer.

Results:

Height of Stairs =			
Volunteer	Weight (Newtons)	Time (Seconds)	Horsepower

To calculate horsepower, use the following formula:

HP = wt. of volunteer (Newtons) x ht. of the stairs (m)/550 x volunteer's time (sec)

1. Could you produce twice as much horsepower by running up some stairs that are twice as high? How do you know?

2. If you did this experiment with a ramp instead of stairs, would the length of the ramp make any difference in the amount of horsepower produced? Why?

3. How long can volunteers continue to produce horsepower? What happens?

4. Can girls produce as much horsepower as boys can? Justify your answer.

5. Trace the energy transformations from the volunteers (where did they get their energy) to the horsepower they generated.

Zipped or Zapped?

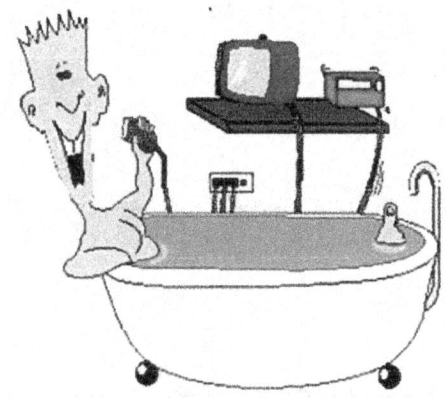

Objectives:
    Students will identify elements by counting the numbers of neutrons and protons in the atoms.
    Students will predict radioactivity of an element.
    Students will manipulate models.

Materials:
    Petri dishes, pinto beans (protons), black-eyed peas (neutrons), periodic table of elements, map pencil, paper plate.

Procedure:
1. The petri dish "atoms" are arranged at lab stations. They are numbered and represent nuclei of isotopes of elements and of atoms. Pinto beans represent "protons", and black-eyed peas represent "neutrons." Some of the containers are "radioactive," (look for the radioactive symbol beside the container number). Other containers are labeled "stable" (this means they are not radioactive).
2. Count the beans and peas in each petri dish. Record these numbers in the data table. **DO NOT OPEN THE CONTAINERS.**
3. Identify each using the periodic table and record in this information in the data table.
4. Calculate the ratio of neutrons to protons and record this in the data table.
        EXAMPLE -- # of beans = 8, # of peas = 12
        neutrons divided by protons = peas/beans = 12/8 = 1.5.
5. As you move from lab station to lab station repeat steps 1 through 4 at each station. (Be sure to record "stable" or "radioactive" clues found on containers in your data table.
6. After you have filled in the data table, use the ratios to determine if an isotope is likely to be "radioactive."
7. Predict the stability for the unknown element. (This container may be opened and poured onto the paper plate for counting).

Results:
1. What sub-atomic particle in each of the various elements does not vary?

2. What sub-atomic particle can vary?

3. Which of the sub-atomic particles helps you identify the element on the periodic table?

4. What is common to all the stable isotopes?

5. What do you know about the "unknown" element?

6. Why must models be used to study atoms?

Zipped or Zapped Data Table

Container	Element or Isotope	Neutrons (peas)	Protons (beans)	Ratio N/P	Stable or Radioactive
A-1					S
A-2					S
A-3					R
B-1					R
B-2					S
B-3					
B-4					
C-1					R
C-2					
C-3					R
C-4					R
D-1					
D-2					S
D-3					
D-4					R
E-1					R
E-2					
E-3					S
E-4					R
F-1					
F-2					
F-3					S
F-4					
Unknown					

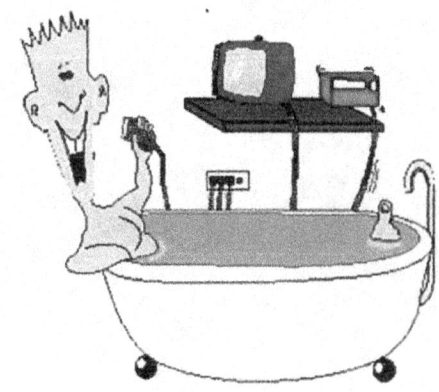

Zipped or Zapped?
Teacher Instructions

Prepare 24 petri dishes with the following information and contents. Copy * items to petri dishes.

Label*	Element	Peas Neutrons	Beans Protons	Stable or Radioactive
A-1	Hydrogen (H-1)	0	1	S*
A-2	Hydrogen (H-2)	1	1	S*
A-3	Hydrogen (H-3)	2	1	R*
B-1	Helium (He-3)	1	2	R*
B-2	Helium (He-4)	2	2	S*
B-3	Helium (He-5)	4	2	R
B-4	Helium (He-6)	6	2	R
C-1	Carbon (C-10)	4	6	R*
C-2	Carbon (C-12)	6	6	S
C-3	Carbon (C-13)	7	6	R*
C-4	Carbon (C-14)	8	6	R*
D-1	Nitrogen (N-12)	5	7	R
D-2	Nitrogen (N-14)	7	7	S*
D-3	Nitrogen (N-15)	8	7	S
D-4	Nitrogen (N-16)	9	7	R*
E-1	Oxygen (O-15)	7	8	R*
E-2	Oxygen (O-16)	8	8	S
E-3	Oxygen (O-18)	10	8	S*
E-4	Oxygen (O-19)	11	8	R*
F-1	Neon (Ne-19)	9	10	R
F-2	Neon (Ne-20)	10	10	S
F-3	Neon (Ne-22)	12	10	S*
F-4	Neon (Ne-23)	13	10	R
Unknown	Uranium (U-235)	143	92	R

You may wish to seal the petri dishes with tape to prevent the students opening them (except the unknown). Labeling the petri dishes "A-1" through "Unknown" can be done with a grease pencil or by using a small piece of masking tape on which is written the identifying information. Be sure to indicate on those marked "*" the term "stable" or "radioactive". You may wish to replace the tern "Radioactive" with the radiation symbol.

200

Index

www.ingramcontent.com/pod-product-compliance
Lightning Source LLC
Chambersburg PA
CBHW081721220526
45468CB00008B/1932